New Horizons in the Philosophy of Science

Edited by

DAVID LAMB

Department of Philosophy
University of Manchester

Avebury

Aldershot · Brookfield USA · Hong Kong · Singapore · Sydney

© David Lamb, 1992

All rights reserved. No part of this publication may be reproduced, stored in a retrieval system, or transmitted in any form or by any means, electronic, mechanical, photocopying, recording or otherwise without the prior permission of the publisher.

Published by
Avebury
Ashgate Publishing Limited
Gower House
Croft Road
Aldershot
Hants GU11 3HR
England

Ashgate Publishing Company
Old Post Road
Brookfield
Vermont 05036
USA

A CIP catalogoue record for this book is available from the British Library and the US Library of Congress.

ISBN 1 85628 296 1

Printed and Bound in Great Britain by
Athenaeum Press Ltd., Newcastle upon Tyne.

Contents

Acknowledgements   vii

1 Introduction   1
  David Lamb

2 Towards a Critical Philosophy of Science   4
  Richard F. Kitchener

3 Philosophy and Frontier Science: 26
  Is There a New Paradigm in the Making?
  Mark B. Woodhouse

4 The Recent Case Against Physicalist Theories   49
  of Mind: A Review Essay
  Joseph Wayne Smith

5 Physics and Existentialist Phenomenology   66
  Robert C. Trundle

6 The Evolution of Science and the   87
  'Tree of Knowledge'
  Jean Marie Trouvé

7	A Sociological Perspective on Disease *Kevin White*	115
8	Sociobiology, Ethics and Human Nature *Lucy Frith*	127
9	The 'Evolutionary Paradigm' and Constructional Biology *Brian Goodwin* *Gerry Webster* *Joseph Wayne Smith*	156
10	Challenge of Ill Health *E.K. Ledermann*	169
11	Death: The Final Frontier *David Lamb*	178
	Notes on authors	187
	Index	190

Acknowledgements

Special thanks to the editors of *Explorations in Knowledge* for permission to use revised versions of papers originally submitted to the journal, especially for Mark B. Woodhouse, 'Philosophy and Frontier Science: Is There a New Paradigm in the Making?', Vol.IV, 2, 1987; Joseph Wayne Smith, 'The Recent Case Against Physicalist Theories of Mind', Vol VI, 1, 1989; Kevin White, 'A Sociological Perspective on Disease', and Brian Goodwin, Gerry Webster and Joseph Wayne Smith, 'The "Evolutionary Paradigm" and "Constructional Biology"', Vol IV, 1, 1987. Thanks also to Josephine Gooderham and Suzanne Evins of Gower publishers for help and encouragement throughout this project and to Mark Mower for technical assistance in the production of the manuscript.

I wish to express my thanks for the support of the University of Manchester Research Fund.

1 Introduction

David Lamb

This collection of ten papers from scholars drawn from Australia, France, the USA and UK all celebrate the diversity and richness of post-positivist philosophy of science. Although wide-ranging in subject matter, with inquiries extending across physics, biology, sociobiology, the social sciences, medicine and the neurophysiological sciences, all of the contributors stress the need to transcend mechanist reductionist science. They also share a commitment to the view that philosophical inquiry can contribute to scientific practice and that philosophers can even take sides in scientific disputes.

The first chapter by Richard F. Kitchener offers a fitting introduction to this collection as it provides a critical survey of positivist philosophy of science and suggestions for new directions. In opposition to positivism and its neglect of metaphysics and ethics Kitchener outlines a philosophy of science which includes epistemology, synthetic metaphysics and axiology of science. This critical approach breaks down artificially imposed barriers between philosophy and the history, sociology, and psychology of science, and outlines steps towards a pluralist nonfoundationalist internal critique.

A critical philosophy of science can do no better than investigate frontier science. In Chapter II, Mark B. Woodhouse considers the relationship between philosophy and frontier science in the context of an examination of theories of non-local causation in physics; holographic information models in neurophysiology; morphogenic fields in biology, and near death experiences in medicine. He concludes his critique of mechanistic explanations with a robust argument for an expanded version of materialism.

Joseph Wayne Smith, in Chapter III, focuses on physicalist theories of mind in an examination of recent challenges to the ruling metaphysical paradigm in the philosophy of mind. Many philosophers not only unquestionably assume an identity between mental events and physical neurological events but also find problems in the very nature of mental and intentional concepts, seeking to eliminate them from scientific discourse. In contrast, Smith challenges the reductionist approach and reveals how a growing body of scientifically literate philosophers are united in their belief that human beings are not *mere* mechanisms and that the human being is not reducible to the human brain.

Robert C. Trundle, in Chapter IV, locates existential phenomenology in the forefront of philosophical inquiry into science. In a unique account of the relationship between theory and observation he maintains that 'the phenomenological consciousness of observation belives the fact that such observation is laden with features which phenomena have *in themselves* independently of observation theorectical concepts.' The subject-matter of his phenomenological analysis in physics.

Historians and sociologists of science have demonstrated how scientific knowledge varies over time and even between different groups of scientists within a particular period of time. Laboratory studies have also revealed how the content of study also varies during the inquiry, and even when results are prepared in a research report or conference paper the contents may vary during the negotiation process. Given the variability and flexibility of scientific knowledge the problem of producing models which reflect scientific activity is daunting. In Chapter V Jean Marie Trouvé takes up this challenge and offers a 'tree of knowledge' model with radical implications for our understanding of science and the growth of knowledge.

In Chapter VI Kevin White explores different methods of approaching the concept of disease. He identifies three concepts: Cartesian (empirical), Hegelian (normative) and Nietzschian (social). The first two are rejected on the grounds that they are based on ahistoric and absolutist concepts of disease and White consequently argues the case for a Nietzschian account according to which diseases are social phenomena constituted by the society in which they are located.

In Chapter VII Lucy Frith's paper focuses on the controversy generated by the discipline of sociobiology. She points out how sociobiological analysis is employed to shore up various philosophical positions within ethics and debates on human nature. Mounting a twofold critique Frith examines the scientific status of sociobiology and its relevance to broader philosophical questions concerning human nature and ethics. Problems with Darwinian theory are addressed in Chapter VIII by Brian Goodwin, Gerry Webster and Joseph Wayne

Smith. Their thesis concerning 'constructional biology' represents a challenge to Darwinism and its 'Neo Darwinian' synthesis; they outline its failure to address the problem of biological form and the problems of morphogenesis and taxonomy. This failure, so the authors argue, is due to the strong genocentric assumptions of the evolutionary paradigm. Following their critique of Neo-Darwinianism the authors suggest that the solution to the problem of biological form lies in the adoption of an alternative ontology of the organism: a field conception of organicism.

The contribution of E.K. Ledermann in Chapter IX develops a social or holistic alternative to mechanistic reductionism by means of a series of brief case studies. An essential characteristic of Ledermann's approach is his belief that freedom of conscience is realised when the patient actually meets the challenge of ill health. This complex subject is developed fully in his book, *Mental Health and Human Conscience: the True and False Self*, Gower, 1984. The final chapter by David Lamb offers a brief survey of philosophical problems raised by neurological criteria for diagnosing death.

D. Lamb. 1992.

2 Towards a critical philosophy of science

Richard F. Kitchener

Introduction

Up until fairly recently, and still continuing in many quarters, the received view about the philosophy of science was one rooted in Logical Positivism/Logical Empiricism.[1] In fact, it is something of a commonplace to point out that logical positivism both established the philosophy of science as a professional specialty (with its own disciplinary identity) and in so doing left an indelible philosophical mark upon it. Indeed, as Suppes points out,[2] although logical positivism was widely rejected as a general epistemology, many continued to see it as an adequate philosophy of science. Logical positivism may have died as a general epistemological or philosophical movement, therefore, but it was merely transformed into philosophy of science. Indeed, although many philosophers of science would argue that this positivist philosophy of science is inadequate, its characteristic orientation (e.g., 'the logic of science') still largely dominates the field of philosophy of science.

During the heyday of logical positivism (and to a lesser extent even today), philosophy of science was identified (roughly) as what logical empiricists such as Carnap, Hempel, Reichenbach, and Feigl and those sympathetic to them (e.g., Nagel, Braithwaite, Salmon) did in the way of articles and books on the philosophy of science. It was never admitted that there were full-fledged alternative and competing philosophies of science - since this possibility would have seemed to make philosophy of science 'non- scientific'[3] - or if this was admitted, these alternatives were seen as family disputes between, say, Carnap and Reichenbach. What seemed unquestionable, however, was that philosophy of science simply was analytic philosophy of science, a point Brodbeck explicitly makes.[4] If one were to suggest there were non-analytic philosophies of science, for example, phenomenological philosophy of science or Whiteheadian philosophy of science, it would not have been difficult to detect a thinly disguised attitude of contempt or amusement among many philosophers of science.[5]

Such a situation remains to a surprisingly large extent still true of the philosophy of science today. If one peruses the pages of the two most prestigious journals devoted explicitly and exclusively to the philosophy of science - *Philosophy of Science* and *British Journal for the Philosophy of Science* - one will look mostly in vain for articles dealing with the philosophy of science of Husserl,[6] Heidegger, Dewey, Whitehead, Polanyi, Piaget, Lorenzen, Brunschvicg, etc.[7] This is not because no one is writing articles on these figures - they are - but rather because these journals do not publish articles of this type. In short, if we label the kind of philosophy of science done by the logical empiricists and contemporary Anglo-American philosophers of science as 'analytic,' then the point I am making is that contemporary philosophy of science is rather exclusively analytic, and non-analytic philosophy of science is ignored or derided.[8]

Science and philosophy of science

According to the received view regarding the philosophy of science, philosophy and science are sharply to be separated. Since there were only two kinds of cognitively meaningful assertions - those of logic and mathematics (on the one hand) and those of the empirical sciences (on the other) - and since there was a sharp distinction made between the analytic/synthetic, the a priori/a posteriori, and the necessary/contingent - the philosophy of science could either be construed as being empirical (and hence like the empirical sciences) or as logico-mathematical (and hence like logic and mathematics). The first option would have produced a 'naturalistic' philosophy of science[9] - something that Otto Neurath seems to have suggested early on - or a 'logistic' philosophy of science,[10] one that turned philosophy (and philosophy of science) into a kind of applied logic.[11] With the influence of Russell, Frege and Bolzano being so strong on the Vienna Circle, it can hardly be surprising that under the particular influence of Carnap, the early positivists became logicists and Platonists.

According to this standard logicistic view, philosophy of science did not make any kind of empirical assertions (which was the proper task of science), but rather its propositions were analogous to (or reducible to) those of logic and mathematics. Furthermore, since logic and mathematics were taken to be analytic, a priori and necessary,[12] so philosophy of science was construed to be analytic, a priori and necessary, whereas science was synthetic, a posteriori, and contingent. As a second order reflective discipline, philosophy of science would never compete with science - and this meant it would make no empirical claims that might conflict with science - but would be a meta-level 'analysis' of science, and analysis, being assimilated to applied logic, would possess those features of logic - its analyticity, a priori nature and necessity - that would make it categorically distinct from the empirical realm.[13]

Thus, on this view, philosophy of science was concerned exclusively with logical analysis. In the inaugural issue of *Philosophy of Science*,[14] Carnap asserted that philosophy of science deals with 'the logical analysis of the concepts, propositions, proofs, theories of science, as well as ...[its]... methods.' As P.H. Nidditch[15] puts it:

> the dominant conception of the nature of philosophy among philosophers in the English-speaking world [is that] philosophical problems are not empirical, and philosophical utterances ... are not empirical: their non-empirical character is bound up with their being in some special sense 'explanatory'. This conception arose by virtue of the desire not to fall into metaphysics on the one hand nor to armchair science on the other.

Furthermore, just as (applied) logic and meta-mathematics are activities to be judged by their own intrinsic (logical or formal) standards, so philosophy of science was imagined to be an activity carried out according to autonomous philosophical (i.e., logical) methods and judged by criteria of a purely philosophical (i.e., logical) kind. This meant that the philosophical correctness of a proffered philosophical analysis of, say, some scientific concept was not to be evaluated by reference to actual scientific practice or its history; this would be tantamount to confusing philosophy of science with science itself. Thus, philosophy of science was not descriptive of actual science, but rather prescriptive of it, or (as they preferred to express it) philosophy of science offers *explications*[16] of scientific concepts and *rational reconstructions* of scientific activity.[17] Furthermore, philosophy of science was radically different from the apparently related disciplines of history of science, psychology of science, and sociology of science. Nothing in these latter (empirical) sciences could be of any relevance to the philosophy of science, since this would be confusing philosophy of science with science[18] and would involve a fallacy of deriving a norm from a fact.

As a consequence of their heavy reliance upon the views of Russell, Frege and Bolzano, who were largely Platonistic in their philosophy of logic and mathematics, most of the early logical positivists were logicists about the foundations of mathematics. In their philosophy of science, this logicistic method resulted in another kind of 'logicism', namely, the thesis (roughly put) that all significant philosophical questions were reducible to questions of logic.

What was crucial to science, therefore, was not scientific activity itself - what we can call the entire scientific enterprise - but rather a set of scientific propositions. Secondly, it was the underlying logic of these propositions that was crucial and this meant, in keeping with the positivist's interpretation of logic, that their *formal* aspect and not their *content* was what was important. To obtain this, one had to abstract the formal aspect from the wider context of scientific thought and canonize it via some abstract model. Initially, this took the form of maintaining, as Carnap did, that what was significant about scientific knowledge could be completely captured by the logical syntax of language. Later, Carnap's restrictive proposal was broadened so as to include the semantic aspect of logic (and sometimes it was even admitted that logic had a pragmatic aspect).[19] The result of such a philosophy of science was a philosopher's philosophy of science not a scientist's. This often resulted in a philosophy of science fiction rather than a philosophy of real science. When this occurred, the 'analyses' were of little relevance to science but ones which would allow generations of graduate students to pursue dissertation topics on, say, the paradox of the ravens or the 'grue- bleen' paradox (to mention just two). Anyone who has ever taught Goodman's 'new riddle of induction' to scientists will recall immediately the struggles involved in convincing them that this has anything whatsoever to do with questions about empirical evidence.

This view of the connection between science and philosophy, together with the associated view of the nature of the philosophy of science, has fewer followers today than it once did (although I think the majority of philosophers of science still hold something like this received view - just look at the kind of articles they typically publish). The reasons for this change are complex and have been documented historically by several individuals. But the upshot is that it is no longer defensible to hold this view of the relation between philosophy of science and science. Philosophy of science ought to be and increasingly is a philosophy of real science: it is concerned with understanding science philosophically, and anything that can help us do this should be used. This includes the history of science, the psychology of science, and the sociology of science. In particular, philosophical analyses of science need to be checked against real science, since a constraint on any adequate philosophy of science is that it match actual cases of science at their paradigmatic best.[20]

That the philosophy of science cannot be sharply separated from the history of science now seems to be widely accepted by philosophers of science. Furthermore, many of them argue that a logicist conception of the philosophy of science must give way to a more historicist conception of the philosophy of science and that a stress on formalism must be balanced by a stress on the development of scientific theories. This would involve, as I have argued elsewhere,[21] seeing them as developmental entities which change according to certain developmental dynamics. Furthermore, to understand science, one must understand real scientific practice with all its complexities and messiness and this, in turn, would involve one in the sociology of science. If one is going to understand how actual science is practiced today, then a description and explanation of that practice is essential. This would include studying science as a social institution, understanding the social norms operating to reward scientific merit and creativity and punish fraud and deception.[22] To understand science we must understand how graduate students are educated, the role of textbooks in the entire scientific enterprise, the need to publish articles in order to receive tenure and promotion, the dissemination of knowledge through conferences and informal groups, the role of the Nobel Prize in directing scientific practice, the epistemological importance of the 'research unit' vs. the individual scientist, etc. If these topics do not appear relevant to the understanding of science, then just consider the absolutely crucial but philosophically neglected role of funding and grants in contemporary science (and its associated political influence on science). Anyone who knows a smidgeon about current science knows the absolutely dominating role that grants play in the scientific endeavor and yet one will look in vain for a philosophical discussion of this feature of science. Instead, what one finds are discussions of, say, Popper on why scientists (should) abandon a theory when it is falsified (via *modus tollens*). True, there are numerous discussions in the sociology of science which are of questionable value (e.g., citation counting) and others (such as the 'Strong Programme') which are philosophically questionable. But there are also numerous other parts which are positively valuable. Although the psychology of science is a discipline that remains to be developed, it too would go far towards helping us understand science in the modern world. I think that a more adequate rendering of the philosophy of science should be conceived in which one includes the history, sociology and psychology of science in addition to the philosophy of science.

The resistance of the positivists to such a view was philosophically anchored in the questionable assumption that one can separate (in a rigidly compatmentalized way) empirical science from philosophy. At most, however, there is a difference in degree between good science and good philosophy, since both must involve conceptual analysis and empirical facts. Furthermore, if one is going to pay attention to real science, one must know something about it and this means that a philosopher of science must not only read an occasional textbook on, say, physics, but must have considerable theoretical and practical experience in the science about which (s)he is philosophizing. The extent of this knowledge may even have to include the conducting of experiments and/or the making of observations as well as a detailed knowledge of the appropriate advanced mathematics that is closely related to the science in question.

If this is correct, then I think a more critical philosophy of science would include the philosophical perspective that science and philosophy are not sharply separated, that the philosophy of science must be rooted in a thorough knowledge of the relevant sciences and even that in order to answer a 'philosophical' question, it may be appropriate to fund research of a purely 'scientific' kind, say, conducting a research program on a particular phenomenon such as the Kervran effect. In short, a more correct philosophy of science should definitely be a scientist's philosophy of science and not a philosopher's philosophy of science.

Metaphysics and science

Philosophy of science on the received view is concerned with the logic of science. It

> interprets science as a body of deductive or quasi-deductive systems of assertions: these systems and their components are analyzed and judged by using concepts and rules (e.g., modus tollens) belonging to the study of formal inference and implications, or to some extension of this, e.g. the probability calculus, or are analyzed into something modelled on it, e.g. a formal theory of simplicity.[23]

Thus, the scope of philosophy of science was basically that of logic - a kind of applied logic. At least, methodologically speaking, philosophy of science was pursued by means of methods basically similar to those found in logic. What then of other fields of traditional philosophy, namely epistemology, metaphysics and ethics? Clearly, epistemology was to be included, since science obviously consists (on the received view) of a set of propositions (claims) about knowledge of the natural world. Scientists not only construct theories, laws, and models, which have an underlying logic to them and whose logical analysis was the task of philosophy of science, they also make various kinds of epistemic claims about these entities and engage in epistemic activities (e.g., observations and reasoning), which are designed to support these claims. Thus, philosophy of science came to be seen as dealing with logic, methodology and epistemology of science.

On this view, therefore, logic and epistemology have a crucial part to play in the philosophy of science but other areas of traditional philosophy have less clear ones. This is especially true of ethics and (to a lesser extent) metaphysics. As Herbert Feigl puts it:

> [Philosophy of science] may involve reflections upon problems traditionally classified as 'metaphysical' ... But the present trend favors a restriction of the discipline of philosophy of science to the logical analysis and clarification of the knowledge-claims of the sciences.[24]

The status of the metaphysics of science in logical positivism has always been somewhat unclear. On the one hand, metaphysics (and ethics) were declared to be cognitively meaningless: 'In the domain of metaphysics', Carnap says,[25] 'including all philosophy of value and normative theory, logical analysis yields the negative result that the alleged statements in this domain are entirely meaningless.' Metaphysical statements, Carnap claimed, were simply *Lebenseinstellungen* and *Lebensgefühle*, subjective existential commitments and expressions of personal feeling.

As such, metaphysical statements did not belong, properly speaking, to science or to the philosophy of science,[26] since science included only what was cognitively meaningful. Obviously, a key question here concerns the meaning of 'metaphysics'. According to Carnap, metaphysical statements 'transcend the limits of human knowledge' in the sense that there is no empirical method for verifying such statements. This, in turn, meant (roughly put) that from such metaphysical statements, no protocol or observation statements were deducible such that these observation statements would verify, falsify or confirm the truth of the metaphysical statement. Thus, by 'metaphysics' Carnap meant

> any alleged knowledge by pure thinking or by pure intuition that pretends to be able to do without experience. But the verdict equally applies to the kind of metaphysics which, starting from experience, wants to acquire knowledge about that which transcends experience by means of special inferences (e.g. the neo-vitalist thesis of the directive presence of an 'entelechy' in organic processes, which supposedly cannot be understood in terms of physics: the question concerning the 'essence of causality,' transcending the ascertainment of certain regularities of succession; the talk about the 'thing in itself'.[27]

In his afterward to this article (written in 1957) Carnap says that the term metaphysics (appearing in his earlier article) refers to the field of alleged knowledge of the essence of things which transcend the realm of empirically founded, inductive science. Metaphysics, in this sense, includes systems like those of Fichte, Schelling, Hegel, Bergson, and Heidegger. But it does not include endeavors towards a synthesis and generalization of the results of the various sciences.[28] Indeed, following Feigl, we can call this latter type of metaphysics an *inductive metaphysics*, which designates 'speculative extrapolations based on scientifically obtainable evidence',[29] e.g., cosmology and psychoanalysis. This may be risky, Feigl says, but not meaningless.

Feigl goes on to allow still another kind of metaphysics as legitimate - *categorical analysis*, which is 'an investigation of the basic concepts and conceptual frames used in our knowledge of reality. This is not fundamentally different from the sort of logical analysis pursued by the positivists.[30] Thus, an investigation into the basic concepts of, say, physics - matter, energy, space, time, causality, determinism - would presumably be an example of this kind of metaphysics. Indeed in another place Feigl explicitly calls such reflections 'metaphysical' and allocates this task to the philosophy of science (even though it is overshadowed, Feigl says, by investigations into the epistemology of science).[31] Likewise, Carnap's[32] famous distinction between internal questions and external questions would presumably be still another example of this third type of legitimate metaphysics.

In addition, therefore, to a *transcendent metaphysics,* which the positivists found objectionable, they were prepared to allow for the legitimacy of an *inductive metaphysics*[33] - or perhaps even a *hypothetico-deductive metaphysics*[34] - and Feigl's categorical analysis. Both of the latter 'legitimate' types of metaphysics might also be called a *philosophy of nature*, i.e., an investigation into 'the nature of nature.' However, what the philosophy of nature is supposed to include, and how it differs both from science and the philosophy of science are questions difficult to answer.

On the one hand, a legitimate philosophy of nature should not be what is often caricatured as 19th Century *Naturphilosophie*, a scientifically irresponsible philosophical anticipation of science in which philosophers purport to advance some kind of transcendent or transcendental supra-scientific knowledge of nature. But, on the other hand, is the philosophy of nature anything more than a coherent synthesis of our best scientific theories and hence not different in principle from science itself? Is there something distinctive a philosopher can contribute towards the philosophy of nature? Is there perhaps a distinctive kind of philosophical knowledge of nature?[35] If so, what is its nature?

Secondly, how is a philosophy of nature different from philosophy of science? Although Compton[36] has argued that the philosophy of nature is different from the philosophy of science, his views seem to me to be questionable. For what he apparently means is that philosophy of science deals with the logic of inquiry, the structure of explanation and the analysis and clarification of concepts, whereas the philosophy of nature inquires into 'what it is as such to be an event or an entity in nature, possessing certain qualities and enjoying relations and changes.'[37] Here Compton seems to be adopting the old positivist's conception of the philosophy of science as the logic, methodology and epistemology of science. But on the rather unexciting claim I am advancing, philosophy of science should also include the ontology of science and hence philosophy of nature belongs to the philosophy of science.

Philosophy of nature seems to be largely concerned with what we can call *the ontology of nature* and with a synthetic view of nature rather than with an analytic one. But should philosophy of nature be seen as anything else than a correctly interpreted philosophy of science, in particular, as the metaphysics of science? According to Taylor,[38] for example, philosophy of nature is a kind of cosmology - the investigation of the most general characteristics of external nature. As such it deals with the nature of matter, mechanism vs. teleology, space-time, evolution, etc. But, as I have already pointed out, such questions as these are paradigmatic examples of issues discussed in the philosophy of science and belong squarely in the positivistic tradition. In short, philosophy of science should include the metaphysics (or ontology) of science as a fundamental part and should not be construed merely as the epistemology of science.

Two other ways in which the logical positivists were involved in metaphysics, which I do not have time to discuss, include the realism vs. non-realism issue about the ontological status of theoretical entities and the distinctive positive metaphysics of positivism - the philosophy of logical atomism - underlying their entire philosophical program.

With regard to the first issue - scientific realism - it was (not surprisingly) those philosophers of science who have been most explicitly committed to realism (e.g., Feigl, Smart, Sellars, Harre, and Popper), who have advanced ontological claims about, say, materialism, whereas phenomenalists, instrumentalists and conventionalists have wanted to avoid all kinds of scientific metaphysics. Thus, if one thinks that science can advance something approaching the correct view of reality, (s)he will be attracted toward the position of realism about the ontological status of theoretical entities, whereas those who deny this will tend to be non-realists. It is no accident that those who, like Duhem, want religion to give the ultimate view of reality would remove this ontological function from science, nor is it an accident that phenomenologists (such as Husserl and Heidegger), pragmatists (such as Dewey) and ordinary language philosophers (such as Ryle and Wittgenstein) have been anti-realists. For all of these thinkers, human experience (as lived through) is the ultimate criterion of what is real and scientific theories are ideal abstractions from this *Lebenswelt*. Classical positivism (e.g., Mach and Pearson) is a philosophical bedfellow with these other apparently radically different philosophical movements, but they were all united by a common rejection of scientific realism. Hence, technical issues in the philosophy of science concerning the ontological status of theoretical entities do make an important difference to one's view about 'metaphysics and science.'

That the underlying metaphysics of logical positivism was logical atomism and that this was a full-blown ontology is a point that hardly is new but for our purposes one worth reiterating.[39] This ontology was not one inspired by a careful study of science - as one might expect - but rather was a priori in nature, motivated by purely logic- philosophical considerations. Indeed it seems undeniable that logical positivists must have done some serious compartmentalization in order to adhere to a logical atomism (on the one hand) while at the same time investigating the logic of field theory, relativity theory, quantum mechanics, etc. in physics. Here were philosophers of science, well-acquainted with the revolutionary developments in physics from 1905 to 1927, who simply failed to incorporate any of these 'new views of reality' into their metaphysics at all; instead, they continued to adhere basically to an ontology based upon Newtonian Mechanics.[40] Why? I can only suggest two reasons. One, they were (initially) anti-realists and did not believe the new sciences were giving them a new ontology, especially when it corrected their view that their experience - interpreted positivistically - was the only reality. Second, they were conventionalists about metaphysical commitments: the issue of one metaphysical view versus another, say, idealism versus realism, was seen to be cognitively meaningless, and, simply put, were personal matters of taste, subjective preferences, existential commitments, pragmatic decisions, etc. This view emerges most clearly in Rudolph Carnap's 'Empiricism, Semantics and Ontology' but is even present in his earlier work *The Logical Structure of the World*.[41] Carnap's philosophy of science has always had a strong commitment to conventionalism - a point still not sufficiently appreciated in contemporary discussions of logical positivism.

After having said that the logical positivists did give some role to metaphysics in the philosophy of science, I have to add quickly that I do not think this ever equalled the importance they attached to the epistemology of science and that, consequently, the place of a metaphysics of science was underestimated. This situation has largely continued throughout the philosophy of science, say, from 1950 to the present. Today, philosophy of science is still basically concerned with the logic, methodology and epistemology of science. True, there are numerous discussions of realism vs. anti-realism, but this is basically seen as an epistemological issue. There are also numerous discussions of space, time, matter, causality, etc. But what seems lacking are discussions about the implications of these concepts, as contained in theories of physics, for our view of the world. This, in turn, is due largely to the fact that, typically, a single isolated concept, say, matter, is thoroughly discussed but there is a conspicuous absence of discussions about its relevance for more global questions in which an entire metaphysical view of nature is articulated under a single motif and all the various elements integrated into a coherent and unified whole. There is, in short (to use Broad's distinction) plenty of discussion of 'critical' or analytic metaphysics but virtually none of 'speculative' or synthetic metaphysics. That is precisely what is missing from the pages of *Philosophy of Science* and *British Journal for the Philosophy of Science* and that is precisely what is, by contrast, offered in a slate of recent books on 'the metaphysics of contemporary physics.'[42] Insofar as logical positivism (and the type of contemporary philosophy of science inspired by it) does own up to a metaphysics, it is an *analytic metaphysics* as opposed to a *synthetic metaphysics*. As one person recently characterizes the former: 'The basic idea of analytical metaphysics is that the world can be understood by breaking it down into its most fundamental parts, or "constituents".'[43]

Thus, what I am suggesting is that it is not sufficient for a critical philosophy of science merely to allow a place for metaphysics in the philosophy of science; instead a more proper conception of the role of metaphysics must be sufficiently broad to include synthetic metaphysics as well as analytic metaphysics. Up to now, this has been missing from the the mainstream of philosophy of science but it is a direction in which a more critical science of philosophy ought to go.[43a]

Ethics and science

Ethics, insofar as it is interpreted in a non-naturalistic way, is cognitively meaningless according to the logical positivists. Ethical assertions make no claims to be true; they make no claims about what is the case; they are not rationally justifiable. On all these counts they are opposed to factual statements. Facts and values are thus radically different for the positivist and one ought never to commit the naturalistic fallacy of moving from an 'is' to an 'ought'.

On the 'standard' positivist conception of ethics - the 'emotivist', 'imperative' (or what Carnap preferred to call the optative approach[44]) ethical claims (beliefs) were really not beliefs at all but attitudes. Basically they were seen as expressions of existential commitments to a *Lebenseinstellung*, to a rule or a way of life. One could, of course, champion a way of life, but one should not be misled about its nature: at rock bottom these ethical statements were matters of personal taste not subject to rational justification; furthermore, since they had no epistemic standing and since facts and norms are categorically distinct, facts could not be relevant to their epistemic standing. Although Carnap never explicitly noted this, his metaphysical outlook embodied in his classic paper, 'Empiricism, Semantics and Ontology', has considerable bearing on his theory of values.

Internal and external questions differ radically for Carnap. Internal questions can only be answered within the respective linguistic framework - either by empirical or logical means. External questions, which are questions about the linguistic framework as a whole, can neither be answered from within a system nor, in fact, answered at all. For such questions do not call for an answer to a theoretical question so much as a practical decision whether to adopt it or not. External questions thus ultimately involve matters of choice and appear to be indistinguishable from existential commitments. Insofar as they are not beliefs but rather attitudes, they appear to be precisely like value judgements in general. They are, Carnap claims, non-cognitive.[45]

Can these framework decisions have any kind of rational justification, e.g., a pragmatic justification or what Feigl calls a 'vindication'? That is to say, can Carnap be interpreted as claiming that although these are practical matters of choice they are nonetheless subject to some kind of rational constraint? For after all (we are inclined to say) some linguistic frameworks are better suited than others for achieving certain kinds of goals.

Here Carnap is uncharacteristically obscure. On the one hand, he admits that 'theoretical knowledge' will influence these non-cognitive decisions. Thus, given that language is intended to be used for communication of factual knowledge, then '[t]he efficiency, fruitfulness, and simplicity of the use of the thing language may be among the decisive factors.'[46] The acceptance of a framework cannot be judged as being either true or false, according to Carnap, because it is not an assertion at all: '[i]t can only be judged as being more or less expedient, fruitful, conductive to the aim for which language is intended.'[47] It thus appears that alternative linguistic frameworks are to be tested by virtue of 'their success or failure in practical use.' Such testing, Carnap assures us, it essential for scientific progress and those with no useful function will sooner or later be eliminated.[48]

It thus appears that Carnap is offering us the kind of pragmatic justification, which Feigl calls a vindication,[49] and hence there is a rational basis for a framework decision, one rooted in pragmatic success. This is really an illusion, however, for questions about pragmatic usefulness (means-end justifications) are empirical questions about what actually results in what; these facts may *motivate* decisions, Carnap claims, but they cannot rationally justify them.

> The thing language in the customary form works indeed with a high degree of efficiency for most purposes of everyday life. This is a matter of fact, based upon the content of our experiences. However, it would be wrong to describe this situation by saying: 'The fact of the efficiency of the thing language is confirming evidence for the reality of the thing world.'[50]

'Judgments of this kind', Carnap says, 'supply the motivation for the decisions of accepting or rejecting the kind of entities.'[51]

It begins to appear as if Carnap is here advancing views strongly resembling those of Thomas Kuhn,[52] concerning the employment of criteria for theory choice. Kuhn's ideas have been violently attacked by numerous philosophers of science upholding the banners of rationality and objectivity. But it seems difficult to avoid the conclusion that Carnap is not only a conventionalist about theory choice but also a relativist, irrationalist and subjectivist.

In Carnap's only major discussion of ethics per se,[53] the connection between ethical attitudes and external questions is once again not explicitly discussed. Many of his comments, however, are relevant to this question. Carnap's theory of optatives, as he calls it, contains two key theses. (1) *First, the thesis of non-cognitivism:* If a statement on values or valuations is interpreted neither as factual nor as analytic, then it is non-cognitive.[54]

(2) Second, *the thesis of pure optatives:* there are pure optatives, i.e., there are statements with pure optative meaning[55] logically implying no factual claims. It follows that pure optatives are clearly non-cognitive. In a philosophical analysis of a value assertion, it will often turn out, Carnap claims, that there is a pure optative mixed with impure optatives and pure factual statements. But Carnap's point is that all truly ethical statements are analyzable as pure optatives (and hence

without any factual consequences) and that impure optatives will contain both a factual component and a pure optative. Ethical statements, like commitments to linguistic framework, are attitudes, preferences, choices or wishes and are non-cognitive. Like external questions, they call for a commitment, an existential leap of faith. Commitments to a framework, we will recall, cannot be rationally justified but can be causally motivated by 'reasons.' Presumably, therefore, the same would apply to pure optatives, i.e., one can give no reasons for them, but one can cite their causes. On this point Carnap is obscure, for (unlike his earlier discussion) he now seems to admit that there are reasons and not merely causes for an ethical attitude,[56] and hence that the latter can be rationally justified. When someone advances a reason (i.e., a belief) for an attitude, (s)he is claiming the former is a valid reason for the latter, and we must investigate 'whether his belief was obtained in a rational way, i.e., whether it is supported by the evidence available to [him], and ... whether the belief constitutes a rational reason for the preference.'[57] In contradistinction to his earlier claims about external questions being non-cognitive and in contradistinction to his claim that ethical statements are non-cognitive, one can easily interpret his above remarks in such a manner that he appears to be claiming that there are (or can be) reasons for an attitude, i.e., as claiming not only that there are causes of attitudes but also (sometimes at least) reasons. In this case (presumably), the attitude would be justified and hence, we might naturally suppose, not cognitively meaningless.

Unfortunately, this is not the case, for although Carnap admits there can be reasons for an attitude, this is not sufficient to make the attitude cognitive, for any evidence or reasons on the basis of which an attitude is formed is not part of the *meaning* of the statement. On the contrary, the meaning of a statement consists of the factual implications derived or derivable from the statement: 'My thesis of pure optative says merely that there are optatives which do not logically imply any factual statements; the thesis does not say anything about the *reasons* for the attitude expressed in the optative.'[58] Since an ethical statement has no subsequent factual implications, he claims, it has no cognitive meaning and is purely optative. Carnap is prepared to admit that particular facts may be a reason for one's ethical 'beliefs' and that observational data, in turn, may be the reason for one's belief in these prior facts, even though not implied by these beliefs. Such observational data would constitute evidence on which a person's factual belief is founded. Carnap then goes on to say: '... the observational evidence which a person may or may not have for his belief in a statement describing a physical situation, is not part of the meaning of this statement.'[59] Thus, because ethical attitudes logically imply no observational data, they have no meaning, *even though there may be reasons for the ethical attitude*. The meaning of a proposition apparently consists only of the factual propositions it implies and not in any prior evidence in its favor. In short, there may be reasons - even good reasons - for an ethical attitude but it is nonetheless cognitively meaningless!

Carnap's theory of ethics, as a metaethical theory, thus appears to be a non-cognitivist one about meaning but a cognitivist one about the logic of moral reasoning.

Carnap's treatment of ethics is puzzling in still another respect. In an earlier discussion of his views about internal and external questions, I pointed out (without exploring the point) that his views about how external questions are answered bears a strong resemblance to those of Thomas Kuhn's, namely that commitments to certain kinds of scientific theories, frameworks or norms appear to be basically existential commitments lacking any rational basis. In the present context Carnap appears to be saying that such commitments *can* have a rational basis and hence can be justified (although still cognitively meaningless). This suggests, contra his earlier discussion, that external questions - at least ethical ones - can be rationally based. But Carnap proceeds to undercut this interpretation by adopting a position, once more, strongly resembling a relativistic one. Suppose, for example, two individuals A and B disagree about certain ethical norms (or certain scientific theories or 'paradigms'). It is still logically possible (Carnap suggests) for A and B to agree in all beliefs, for their reasoning to be in perfect accord with deductive and inductive canons, and yet for them to differ in their optative attitudes.

> ... both may have exactly the same relevant evidence, apply the same valid inductive method, and thus come to exactly the same degree of credence for all relevant propositions ... The difference between A and B in their decisions ... is based, in this case, not on a difference in their theoretical thinking but rather on a difference in their preferences ... and finally on a difference in character.[60]

Thus, decisions (preferences) are underdetermined by all the rational evidence. If we recall Kuhn's similar views about value judgments in science, we can immediately see that Carnap's position seems to be indistinguishable from it. All we need do is apply the same analysis, *mutatis mutandis*, to theory choice in science to see what Carnap's view must be: two scientists can agree about all the facts and reason impeccably and yet prefer different theories. Hence Carnap appears to be a kind of *fideist* about theory choice in science. Once more, it appears that theory choice in science (as in ethics) must be matters of personal taste - 'de gustibus not disputandum.'

Axiology and science

If scientists qua scientists make value judgments (epistemic or non-epistemic), then such judgments fail to have cognitive meaning and are matters of personal commitment. If, as I have suggested, such a view seems to follow from Carnap's 'reconstructed' theory of values, then one might well understand why the positivists would have spent so little time on the role of ethics and norms in science. For it seems to be clear that, on this meta-normative theory, science would fail to command the kind of allegiance the positivists thought it should (since science would have nothing to say about ethics). Naturally, they would have kept quiet about this aspect of their philosophy of science - or even have repressed it - since it would have endangered their entire philosophical program. It is not surprising, therefore, that positivists systematically failed to discuss 'values and science' and failed to include axiology as a branch of the philosophy of science.

I think it is arguable that the situation is not much better today and the reason for it is that most philosophers of science continue to have a non-cognitivist view of ethics. One indication of this is the appalling lack of discussion of this very issue - values and science. For example, if one consults the *British Journal for the Philosophy of Science,* one will find only one general discussion of this issue in 35 years! If one peruses the pages of *Philosophy of Science*, the situation is not much better: in the last 15 years, there is again only one general discussion (although other articles have appeared in earlier issues). Two articles in 50 years of combined publication is hardly an impressive record, especially since it is now conceded that 'values and science' is a crucial philosophical topic.

This entire situation needs to be redressed. We must sever the ties that bind non-cognitivist ethics and philosophy of science. The philosophy of science must include, as an essential part, the axiology of science.

The epistemic values of science are those values related to the acquisition of (theoretical) knowledge and concern questions of evidence and truth, e.g., simplicity, falsifiability, objectivity, replicability, testability, theoretical coherence, plausibility, power, curve-fitting, alpha level of significance, etc. In addition to epistemic, moral and aesthetic norms associated with theoretical science, the axiology of science should include normative issues as these relate to scientific experimentation and observation (e.g., the treatment of human and animal subjects under experimental conditions) and norms as they relate to applied science and technology (e.g., recombinant DNA, nuclear energy). But, as I have already suggested, to deal adequately with the ethics of science, one must also deal with the hard realities of actual scientific practice. It is clear, for example, that cheating is much more widespread in science than most people - especially scientists - believe it is or will admit. But it does no good to claim scientists ought not to do this if the social institution of science is such that cheating, in some form, seems unavoidable.

A critical philosophy of science

Up to now, I have been suggesting that the philosophy of science should include the epistemology, metaphysics and axiology of science. I have also suggested that such a philosophy of science cannot be sharply separated from science (on the one hand) nor from the history, psychology and sociology of science (on the other). In conclusion, I want to add the following qualification to my conception of the nature and scope of the philosophy of science: it should be a *critical* philosophy of science.

Up to now, philosophers of science have often been guilty of a version of *scientism*, which (broadly conceived) can be characterized as an uncritical, dogmatic worship of science, its methods and results, and a faith in its ability to solve all of our pressing problems. At the very least, it is the view that the scientific way of knowing is the only way, that what is ultimately real is what is contained in our scientific world view, and that what is valuable and how we ought to live is what the latest scientific theory says. I am not saying these views are in fact wrong - although they may well be - I am only saying that there is a need to critically evaluate whether they are true or not. The one thing recent philosophers of science such as Popper and Feyerabend have taught us is the need for a critical stance and this is best served by a theoretical pluralism. In the context of the philosophy of science, this can be put by saying we need competing philosophies of science that engage each other in confrontation; we cannot continue to have the present situation of a monolithic analytic philosophy of science underlying the outlook of professional philosophers of science.

If one is going to be critical of science - in the sense of performing a *critique* of it - one will not simply *describe* what it is, what its epistemology, ontology and ethics are - although this will be an important part. But neither will one *prescribe* - in a philosophically a priori or foundationalist way - what it ought to be (although there will be something of a prescriptive element to it). Either extreme should be rejected as inadequate. I would favor an approach that combines features of both in such a way that ones philosophical 'analyses' or accounts would be tested against science as it really is. How it is to be done is something I cannot go into here.

What I imagine such a 'critique' and a critical philosophy of science to be would be one characterized as an *internal* critique of science, not an *external* one. It would have as one of its goals the *improvement* of science as a result of pointing out some of its current shortcomings. But it would also unmask certain inflated pretensions of science, ones encapsulated in what Bernie Rollin calls the *ideology* of science.[61] As recent philosophers such as Habermas have suggested, one of the tasks of critical approach to science is *ideology critique*, and this would certainly fall within the present conception of the philosophy of science.

Perhaps the best overall summary of what a critical approach to the philosophy of science would include comes from a recent book on Heidegger's philosophy of science.[62] To ask the critical question, one might say (adapting some of Kockelmans' remarks) is:

> to ask the question of precisely what science is, how it is related to all the other orientations of man toward the world, what its prospects and what its limits are, what kinds of contributions the sciences can make to meaningful discourse about religious, moral, aesthetic, social, political, and educational issues, and what are the areas in which, in this regard, one may not expect a positive contribution from the sciences, simply because of the fact that one appears to run into issues which lie far beyond the competence of the scientific method.[63]

One can hardly find a better characterization of what I take a critical philosophy of science to be.

Notes

1 How to distinguish these philosophical positions is difficult to determine but nothing of any importance seems to depend upon it. In the remainder of this paper I will lump all together under the label 'logical positivism.'

2 Frederick Suppes, 'The Search for Philosophical Understanding of Scientific Theories.' *The Structure of Scientific Theories*, 2nd ed., F. Suppes ed. (Urbana: University of Illinois Press, 1977), p.6.

3 This is the impression one gets, for example, from reading the Preface in Feigl and Brodbeck's classic *Readings in the Philosophy of Science* (New York: Appleton-Century-Crofts, 1953). Indeed, one suspects that if there was a 'unity of science' movement within logical positivism, there was also a 'unity of philosophy of science' movement - there is or ought to be just one philosophy of science! One day in class, in response to a question about her philosophy of science, May Brodbeck replied: 'This is not *my* philosophy of science: this is *the* philosophy of science.' Feigl was much more tolerant of diversity.

4 May Brodbeck. 'The Nature and Function of the Philosophy of Science', *Readings in the Philosophy of Science*, H. Feigl & M. Brodbeck, eds. (New York: Appleton-Century-Crofts, 1953), p.5.

5 Cf. Ernan McMullin's ('Recent Work in the Philosophy of Science', *New Scholasticism, XL* [1966], p.509) remark: 'Seldom has a philosophical subject [philosophy of science] been as closely identified with a language [English] and a

geographic location [Anglo-American countries] as philosophy of science has become in this century.'

6 It should be pointed out that this picture has changed somewhat in recent years and that one can now find discussions of phenomenology and the philosophy of science in at least one of these journals. See Patrick Heeland, 'Husserl's Later Philosophy of Science', *Philosophy of Science*, 54 (1987), pp.368-390.

7 Other journals stressing philosophy of science are less doctrinaire but also less prestigious and hence less paradigmatic: *Scientia, Epistemologia, Dialectica, Zeitschrift für allgemeine Wissenschaftstheorie, Explorations in Knowledge, Synthese, Zygon, Inquiry, Journal of the British Society for Phenomenology, Fundamenta Scientiae,* etc.

8 This same point could be put somewhat differently by saying that Anglo-American philosophers of science read virtually nothing except books and articles in English. This even includes analytic philosophy of science done on the Continent. If Stegmüller's work, for example, had not been translated into English, I doubt whether anyone would have read it. Similar remarks apply to Lorenzen's Constructivism and to the Konstanz school. See, for example, Paul Lorenzen, *Konstruktive Wissenschaftstheorie* (Frankfurt: Suhrkamp, 1974); Jürgen Mittelstrass, *Die Möglichkeit von Wissenschaft* (Frankfurt: Suhrkamp, 1974); Peter Janich, Friedrich Kambartel and Jürgen Mittelstrass, *Wissenschaftstheorie als Wissenschaftskritik* (Frankfurt: Aspecte, 1974).

9 For a discussion of one kind of 'naturalistic' philosophy of science see Ronald Giere, *Explaining Science: a Cognitive Approach* (Chicago: University of Chicago Press, 1988) and Paul Thagard, *Computational Philosophy of Science* (Cambridge, MA: MIT Press, 1988).

10 The turn towards logistic philosophy of science seems to have been due largely to the influence of Carnap, with Schlick's view on this question being somewhat less clear.

11 For a discussion of 'logistic' philosophy of science see Ernan McMullin, 'The Ambiguity of "Historicism".' *Current Research in Philosophy of Science*, Peter D. Asquith and Henry E. Kyburg, Jr., eds. (East Lansing, MI: Philosophy of Science Association, 1979).

12 Not everyone in the early Vienna Circle held this view about the radical non-empirical nature of logic. Kurt Gödel, Alfred Tarski and Hans Hahn seem not to have.

13 For example, Carl Hempel ('Aspects of Scientific Explanation', *Aspects of Scientific Explanation and Other Essays in the Philosophy of Science*. [New York: Free Press, 1965], pp.412-15) likens the task of philosophy of science to that of proof theory in meta-mathematics.

14 Rudolph Carnap, 'On the Character of Philosophical Problems', *Philosophy of Science*, 1 (1934), pp.5-19.

15 P.H. Niddich, 'Introduction.' *The Philosophy of Science*, P.H. Niddich, ed. (Oxford: Oxford University Press, 1968), p.5.

16 Rudolph Carnap, *The Logical Foundations of Probability*, 2nd ed. (Chicago: University of Chicago Press, 1962), pp.5-7.

17 Hans Reichenbach, *Experience and Prediction*, (Chicago: University of Chicago Press, 1938), p.6.

18 May Brodbeck, op cit, p.3.

19 For a history of Carnap's changing views about this question see 'Carnap's Intellectual Autobiography', *The Philosophy of Rudolph Carnap*, P.A. Schlipp, ed. (LaSalle. IL: Open Court, 1963), pp.46-56, 53-56, 60-67.

20 I am explicitly ignoring all the philosophical problems underlying this statement. Needless to say, considerably more must be said concerning how this is supposed to work. Lacking such a conception leads to conceptual problems exemplified, for example, by R. Ackermann. *The Philosophy of Science* (New York: Pegasus, 1970).

21 Richard Kitchener, 'Developmental Explanations', *Review of Metaphysics*, 36 (1983), pp.791-818; 'Is Genetic Epistemology Possible?', *British Journal for the Philosophy of Science* 38 (1987), pp.238-299; *Piaget's Theory of Knowledge: Genetic Epistemology and Scientific Reason* (New Haven: Yale University Press, 1986).

22 See Robert Merton, *The Sociology of Science* (Chicago: University of Chicago, 1973).

23 P.H. Niddich, op cit, pp.2-3.

24 Herbert Feigl, 'Philosophy of Science', *Philosophy* (Englewood Cliffs, NJ: Prentice-Hall, 1964), p.470. This same point is made by Robert J. Baum ('Can Governmental Support of Philosophy of Science Research be Justified?', *PSA 1976*. Vol. 1, F. Suppe & P.D. Asquith, eds. [East Lansing, MI: Philosophy of Science Association, 1976], p.293). In fact Baum is one of the few philosophers who have explicitly claimed that philosophy of science should include the epistemology, metaphysics and ethics of science. In fact, he even goes so far as to suggest that every branch of philosophy should be a part of the philosophy of science, including for example, the political philosophy of science, the aesthetics of science, etc.

25 Rudolph Carnap, 'The Elimination of Metaphysics through Logical Analysis of Language', *Logical Positivism*, A.J. Ayer, ed. (Glencoe, IL: Free Press, 1959), pp.60-61. This article originally appeared in 1932.

26 In a sense, however, Carnap's existential or emotive theory of ethics was a kind of 'philosophy of science' as applied to ethics, only it was a philosophy of science in a broader sense than his own conception of the philosophy of science, since the latter was equivalent to 'a philosophy of scientific knowledge' whereas the former was a philosophy of the entire scientific enterprise (including non-epistemic domains).

27 Carnap, *Ibid.*, pp.76-77. Other positivists shared this view. See Moritz Schlick, 'Positivism and Realism', *Logical Positivism*, A.J. Ayer, ed. (Glencoe, IL: Free Press, 1959), pp.83-102; A.J. Ayer, 'Demonstration of the Impossibility of Metaphysics', *A Modern Introduction to Philosophy*, P. Edwards & A. Pap, eds. (New York: Free Press, 1973), pp.761- 63 and A.J. Ayer's *Language, Truth and Logic,* 2nd ed. (New York: Dover, 1946), pp.33-34.

28 R. Carnap, *Ibid.*, p.80.

29 Herbert Feigl, 'The Weiner Kreis in America', *The Intellectual Migration: Europe and American,* 1930-1960, (Cambridge: Harvard University Press, 1969), p.655.

30 *Ibid.*

31 Herbert Feigl, 'Philosophy of Science', p.470.

32 Rudolph Carnap, 'Empiricism, Semantics and Ontology', reprinted in Carnap's *Meaning and Necessity* (Chicago: University of Chicago Press, 1952), pp.205-221.

33 See also Errol Harris, *The Foundations of Metaphysics of Science* (London: George Allen & Unwin, 1965), p.29.

34 A.N. Whitehead, *Process and Reality*, D.R. Griffin D.W. Sherburne, eds. (New York: Free Press, 1978), p.3. The term hypothetico-deductive as applied to Whitehead's metaphysics emerged in philosophical conversation with Don Crosby.

35 Harris, *loc. cit.*, pp.169-172 apparently does not believe so, whereas John Compton ('Reinventing the Philosophy of Nature', *Review of Metaphysics*, 33 [1980], pp.2-28; 'Understanding Science', *Dialectica*, 16 [1963], pp.155-176) does, with Ivor Leclerc ('The Necessity Today of the Philosophy of Nature', *Process Studies* 3 [1973], pp.158-168) being somewhere in the middle.

36 Compton, 'Reinventing the Philosophy of Nature.'

37 Compton, 'Understanding Science', p.169.

38 A.E. Taylor, *Elements of Metaphysics* (New York Barnes & Noble, 1903).

39 The most consistent and relentless contemporary defense of logical atomism is Gustav Bergmann. See his *The Metaphysics of Logical Postivism*, 2nd ed. (Madison: University of Wisconsin Press, 1967). On the connection between logical atomism and philosophy of science see his *Philosophy of Science* (Madison: University of Wisconsin Press, 1957).

40 This is clearest in the case of Bergmann, who quite clearly and explicitly adopts Newtonian mechanics as his model of what science should be and constructs his philosophy of science to limn its underlying features. See his *Philosophy of Science*, Ch. 1. In this regard, contrast the work of Wolfgang Köhler, who created a Gestalt philosophy of science explicitly based upon field theory in physics. See his *Die physikalischen Gestalten in Ruhe und in stationären Zustand* (Erlangen: Weltkreis, 1920).

41 Rudolph Carnap, *The Logical Structure of the World and Pseudoproblems in Philosophy*, Rolf A. George, trans. (Berkeley: University of California Press, 1967). This was originally published in 1928.

42 Perhaps the most widely read of these books is Fritjof Capra's *The Tao of Physics* (New York: Bantam, 1976), with a close second being Gary Kukav's *The Dancing Wu-Li Masters* (New York: Bantam, 1980). There are a score of other books devoted to similar themes.

43 Larry Lee Blackman, 'Introduction', *Classics of Analytical Metaphysics*, (Lanham: MD: University Press of America, 1984), p.xiii.

43a See Richard F. Kitchener (ed.), *The World View of Contemporary Physics: Does it Need a New Metaphysics?* (Albany, NY: S.U.N.Y. Press, 1988). Richard F. Kitchener and Kenneth Freeman (eds.), *The Origin of the Universe: Scientific and Philosophical Perspectives* (Albany, NY: S.U.N.Y. Press, 1991) for examples of such a metaphysics.

44 Whereas Carnap endorsed a version of the emotive theory, Hans Reichenbach (*The Rise of Scientific Philosophy* [Berkeley: University of California Press, 1951], pp.276- 302) adopted an imperative interpretation. Herbert Feigl, by contrast, although flirting somewhat with these non- cognitivist interpretations, came to a basically cognitive interpretation of ethical statements. See his 'Validation and Vindication: an Analysis of the Nature and the Limits of Ethical Arguments', *Readings in Ethical Theory*, W. Sellars & J. Hospers, eds. (New York: Appleton-Century-Crofts, 1952), pp.667-680 and 'De Gustibus Non Disputandum...', *Philosophical Analysis*, M. Black, ed. (Englewood Cliffs, NJ: Prentice Hall, 1950), pp.113-147. Feigl's approach had a strong influence on Paul Taylor, *Normative Discourse* (Englewood Cliffs, NJ: Prentice-Hall, 1961).

45 Carnap, op. cit., pp.208, 215.

46 *Ibid.*, p.208.

47 *Ibid.*, p.214.

48 *Ibid.*, p.221.

49 Herbert Feigl, 'De Gustibus Non Disputandum ...?'

50 Carnap, *op. cit.*, p.208, my emphasis. Carnap goes on to say (p.208): 'we should rather say instead: "This fact makes it advisable to accept the thing language".' But, given what he apparently means - that pragmatic considerations causally motivate but don't rationally justify - this is precisely what he cannot say since facts cannot make anything advisable.

51 Carnap, *op. cit.*, p.214.

52 Thomas Kuhn, 'Objectivity, Value Judgments, and Theory Choice,' The Essential Tension (Chicago: University of Chicago Press, 1977), pp.320-330.

53 Rudolph Carnap, 'Replies and Systematic Expositions', *The Philosophy of Rudolph Carnap*, P. Schlipp, ed. (LaSalle, IL: Open Court, 1963), pp.999-1016: 'Value Judgments.' For an incisive critical evaluation of Carnap's (and Reichenbach's) theory of ethics, see Abraham Kaplan, 'Logical Empiricism and Value Judgments', *The Philosophy of Rudolph Carnap*, P. Schlipp, ed. (LaSalle, IL: Open Court, 1963), pp.827-858.

54 Carnap, *op. cit.*, p.999.

55 The optative meaning of a statement is that kind of meaning common to wishes, proposals, requests, demands, commands, prohibitions, preferences, etc.

56 *Ibid.*, p.1006.

57 *Ibid.*, p.1006.

58 *Ibid.*, p.1007.

59 *Ibid.*

60 *Ibid.*

61 See his book, *Animal Consciousness, Animal Pain, and Scientific Change* (Oxford: Oxford University Press, 1989).

62 Joseph J. Kockelmans, *Heidegger and Science* (Washington, D.C.: Center for Advanced Research in Phenomenology and the University Press of America, 1985).

63 Kockelmans, *op. cit.*, p.3.

3 Philosophy and frontier science: Is there a new paradigm in the making?

Mark B. Woodhouse

Introduction

This essay examines four controversial scientific developments that challenge deeply entrenched theoretical frameworks: (I) non-local causation in physics; (II) holographic information models in neurophysiology; (III) morphogenetic fields in biology, and; (IV) near-death experience in medicine/psychology. In each case I show why the topic should inform discussion of specific philosophical issues, such as simultaneous causation, the distinction between form and matter, or the localization of mental states, and I provide independent philosophical analyses. A major conclusion is that the relevant explanatory models collectively require an expanded materialist ontology in which interlocking fields, forces, and frequencies play roles that, within the context of mechanistic explanation, would be assigned to the configuration and/or momentum of discrete material units, such as an electron or a gene.

Philosophy and frontier science: is there a new paradigm in the making?

Philosophy neither begins nor ends with science, but each discipline can be enriched by mutual interaction along the way. To this end I raise for discussion four controversial developments in science for which the evidence not only justifies continued investigation but also poses challenges for deeply entrenched theoretical frameworks. The topics are (I) non-local causation in physics, (II) holographic information models in neurophysiology, (III) morphogenetic fields in biology, and (IV) near-death experience in medicine/psychology. I have selected these developments for three reasons. First, they individually invite philosophical examination in their own right and should inform to a greater extent the discussion of specific issues in philosophy, especially simultaneous causation, the distinction between form and matter, and the localization of mental states. Second, they involve a comparatively narrow cluster of concerns with the conditions that bear most directly on the acquisition, retention, or transfer of information in ways that appear to violate normal material, spatial, or temporal barriers. Third, the models suggested by the anomalous phenomena in question tend to exhibit an expanded materialist ontology in which interlocking fields, forces, and frequencies play roles that, within the context of mechanistic explanation, would be assigned to the configuration and/or momentum of discrete units of matter, such as an electron or a gene.

In each case I present a brief description of the key issue and of the relevant scientific literature. In surveying opportunities for philosophical involvement, I inject both clarification and criticism. However, since my primary aim in this overview is to uncover a pattern, the ratio of exposition to argument is greater than the analysis of a single topic otherwise would require. This essay, then, is not about the direction or limits of empirical inquiry suggested by the analysis of certain philosophical issues. Rather, it is about developments in four areas of frontier science that suggest a major paradigm shift in the making which carries significant implications for long-standing philosophical issues.

Non-local causation in physics

In 1935 Einstein, Podolsky, and Rosen published a now-famous thought experiment purporting to undermine the completeness of quantum-mechanics by means of what, after the authors, is termed the EPR effect.[1] Of several experimental formats, one suggested by David Bohm utilizes electron spin-states. Suppose we generate a two-particle system of zero spin, i.e., a situation in which the spins of paired particles are always equal and opposite and thus, as related constituents of a single system, cancel each other out. According to quantum mechanics if the spin of one ('A') is up, the spin of the other ('B') must be down. Suppose, further, that A and B, having been in close proximity by virtue of a common origin, are now separated in opposite direction and that A's spin is changed artificially from up to down (or left to right, etc.). Quantum theory requires that B's spin simultaneously change from down to up, even if the particles are widely separated. The apparent causal dependence of B upon A in this situation is the EPR effect.

Einstein believed the EPR effect 'paradoxical,' because it violates the assumption of local causation which states that, necessarily, for any two particles to be connected causally, the time between their respective changes cannot be less than that required for the transmission of light between them. Thus a quantum-mechanical prediction violated a foundation of relativity theory, viz., the absolute speed of light. To accept the EPR effect would be to accept 'non-local' causation -- which Einstein did not!

In 1964 J.S. Bell produced a rigorous mathematical demonstration (with subsequent variations) the import of which was that *if* the statistical predictions of quantum theory are true, a realistic universe is incompatible with the assumption of local causation.[2] That is, any explanation of the EPR effect that assumes both that the particles in question are not causally connected and that their existence and behavior are independent of mind must generate predictions that differ from those of quantum-theory.

In 1972 and in some half-dozen subsequent experiments the relevant predictions of quantum-mechanics were put to critical tests. The results were that quantum theory largely has been vindicated and that the inviolability of local causation has been placed in serious doubt. Bell's prediction was confirmed, i.e., that for quantum-theory to hold up, the state of particle B must depend on an instantaneous and non-chance manner upon what an experimenter does to particle A in a separate region.[3] Henry Stapp, a physicist and key interpreter of Bell-related work, described this work as the most important discovery in the history of physics.

Many physicists predictably have adopted the Copenhagen attitude that, if we cannot get at the natural facts behind the numbers, and if the predictions based upon those numbers continue to be confirmed (which they have), there is no point to speculating about what might 'connect' the particles from behind the scenes. For Bohr, no paradox existed. Others have attempted to get around the problem by reformulating relevant portions of quantum theory--with question-begging results.[4] A small but growing number of physicists are facing the prospects for major conceptual revision. Bell's theorem is unshakable, but where it leads is not clear. In their review of the literature, Clauser and Shimony state:

> Because of the evidence in favor of quantum mechanics from the experiments based on Bell's theorem, we are forced to abandon...a realistic view of the physical world (perhaps an unheard tree falling in the forest makes no sound after all) or else to accept some kind of action-at-a-distance. Either option is radical, and a comprehensive study of their philosophical consequences remains to be made.[5]

A comprehensive study, indeed! To begin, the exclusive disjunction between realism and local causation (which defines much thinking about the implications of Bell's theorem) is quite misleading. On the one hand, local causation would not be preserved by adopting idealism, because to suppose that consciousness somehow (psychokenetically?) instantaneously alters the spin of a distant (nonphysical) particle still involves action-at-a-distance.[6]

On the other hand, no physicist has shown how it is possible to avoid violating the principle of local causation in light of the relevant experimental work without encumbering even more bizarre and problematic consequences. We might preserve the principle, for example, by introducing undetectable patterns of backward time travel. However, this proposal would not entail idealism. Preserving the inviolability of local causation does not appear to be a viable alternative. If so, the issue is not realism vs. locality, but rather what becomes of realism when the principle of local causes is modified in light of the Bell experimental results.

The most straightforward response to this issue is suggested by the following argument. If two particles are correlated instantaneously in a non-chance manner, then the source of their connection must lie outside space-time as defined by the speed of light, i.e., must involve 'superluminal' (faster-than-light) communication. Stapp, for example, argues that, barring certain extremely counterintuitive alternatives, such as those stipulated by many-worlds theorists, superluminal communication is: (a) consistent with quantum mechanics; (b) presupposed by the possibility of EPR-type effects, (c) derivable in part from the indivisibility of Planck's quantum of actions, and; (d) indirectly testable.[7]

Whatever the merits of this line may be, it does not entail the mind-dependence of physical processes. Rather, it calls into question the possibility of the complete knowability of the world. In other words the Bell-related work merely suggests a weaker sense of realism via its stipulation of certain limits to scientific knowledge. We cannot get 'outside' space-time to determine exactly why A and B are so correlated within space-time. Realism may be weakened in this comparatively harmless sense, but idealism is not entailed by non-local causation.

Does the concept of non-local causation embody a contradiction? If we adopt a Humean account of causality that requires both spatial contiguity and temporal succession, then any instantaneous change in which the particles are noncontiguous is not one we could label as 'causal' without contradiction. To avoid the contradiction, we may opt for either (1) not to label the connection as causal or (2) to label it as causal, but reject both spatial contiguity and temporal succession as defining conditions of causal connections.

There are several reasons for rejecting (1) in favor of (2). First, there is no viable alternative to labeling the connection as causal. To suppose that they are connected merely by chance is ruled out both by the formalism of quantum theory and by subsequent Bell-related experimental work. Another alternative, pre-established harmony, for well known reasons appears as an even darker cloud on the horizon (to borrow D. M. Armstrong's depiction of parapsychology where parallels to the EPR effect have not gone unnoticed), than the EPR effect itself.

A second reason for rejecting (1) in favor of (2) is that the EPR effect conforms to our most basic conviction regarding the nature of causality. If a change randomly introduced into the spin state of A consistently is accompanied by (is sufficient for) a change in the spin state of B, then we would take them to be causally connected. (A similar argument for the bare possibility of instantaneous action-at-a-distance is developed by J. L. Mackie, although without reference to the EPR effect.)[8]

How could the EPR effect occur? The specific causal mechanisms are a mystery. The handful of defenders of a superluminal interpretation deny that any 'thing' (such as a tachyon) carries information between the particles. Moreover, field theories associated with accounts of action-at-a-distance at best could provide a connecting medium for the particles but still fail to account for the instantaneous interaction between widely separate regions of the field. The dearth of causal underpinnings, however, should not preclude our taking the EPR effect for what it is, namely, an instance of instantaneous, non-local causation.

Philosophers long have been aware of odd quantum-level goings-on. Why suppose that any of this has relevance for macroscopic phenomena? Stapp argues that with suitable modifications Bell's theorem is applicable to large-scale events, although it is not clear how this would work.[9] Perhaps certain synchronicities in traffic patterns--the kind of thing that excites general systems theorists--would become predictable. Of more relevance is William Condon's apparent discovery of causal asymmetry violations in speaker-hearer communication that parallel in some respects those described by non-local causation at the quantum level. Hearers are recorded on film exhibiting distinctively appropriate and repeatable reactions to certain words as they are being uttered, but *before* they reach their ears. Condon ruled out lip reading and concluded that persons sometimes react subconsciously to information coming before it has arrived.[10]

For those interested in a larger framework in terms of which Bell's theorem and other puzzles in recent physics are worked out, Bohm's recent part scientific and part metaphysical treatise, *Wholeness and the Implicate Order*, makes for stimulating reading. Among its central themes is the claim that various developments in physics, including Bell's theorem, lead us to strong monism, a vision of pervasive 'quantum interconnectedness' suggestive of a doctrine of internal relations. In his words:

> The non-local, non-causal nature of the relationships of elements distant from each other evidently violates the requirements of separateness and independence of fundamental constituents that is basic to any mechanistic approach....Thus, if all actions are in the form of discrete quanta, the interactions between different entities (e.g., electrons) constitute a single structure of indivisible links, so that the entire universe has to be thought of as an unbroken whole. In this whole, each element that we can abstract in thought shows basic properties (wave or particle, etc.) that depend on its

overall environment, in a way that is much more reminiscent of how the organs constituting living beings are related, than it is of how the parts of a machine interact.[11]

Holographic information models in neurophysiogy

Neuropsychologists never have explained successfully how memory storage and retrieval occurs in light of the clinical fact that no matter which sections or, up to a point, how many sections of the cortex are removed or destroyed, recall is not substantially reduced. For example, the destruction of half a brain does not reduce recall by half, much less result in the patient recognizing only half his published works or family. Recall appears to be an all-or-nothing situation so far as the brain is concerned. There is no direct correspondence between loss of memory and loss of functional brain tissue. To be sure, various proteins and chemical changes in some way are involved in recall, but the fifty-year search for an engram (a change in brain cells that marks a memory-trace) has yielded little fruit. Moreover, even if engrams were discovered they are inherently incapable of accounting for what Karl Lashley (who borrowed the term from Hans Driesch) called 'equipotentiality,' i.e., the fact that memories are not spread around simpliciter. Rather, each memory appears to be located in many or even all parts of the cortex.

Neuropsychologist Karl Pribram has argued that only a holographic model of the brain's information processing capacities can account for equipotentiality and other long-standing puzzles of cognitive psychology.[12] By way of explicating his proposal, a brief description of three critically relevant features of holography is necessary.

First, the 3-dimensional holograms we see projected in space are literally interference patterns resulting from the intersection of waves of two beams of coherent light. Gradations of amplitude and frequency in principle make it possible to store billions of 'bits' of information in a single complex interference pattern. The hologram is a literal reconstruction of initial physical features by means of intersecting wave-fronts.

Second, holograms are made possible by the conformity of the interference patterns recorded on a plate (part of which emanates from the initial physical object) and the pattern projected in space to a common mathematical form rigorously describable by Fourier analysis, a type of calculus by which the most complex interference patterns are transformed into their constituent sine waves. Only those patterns that yield to this analysis are in principle holographic. We can read into or out of any hologram mathematical analogs of the most minute gradations of physical detail.

Third, the most important feature of a hologram is that each part of its diffuse interference pattern bears a complete code of the object the form of which is spatially projected. This means that a plate bearing an interference pattern (that appears to us as more or less blank) can be broken into small fragments and *each* piece will reconstruct, although with less clarity, the *entire* image. A quarter of a plate, for example, would still produce the image of an entire human body, not one-fourth of a body.

Since both brains and holograms appear to store information equipotentially, Pribram surmises, then the principles of hologram construction must therefore guide information processing in the brain. The experimental evidence for this theoretical insight is far from conclusive, but it is substantial and, more important, steadily growing. Considerable support exists for viewing major neural sub-systems (auditory, tactile, visual, etc.) as 'frequency analyzers' rather than as mere passive transmitters of discrete sense-data.[13] Direct evidence for wave-fronts of neural firings and interference patterns has been developed by E.R. Johns.[14]

Pribram's own work has centered on the visual system. That cells in the visual cortex appear to respond to geometric form, as he and others have shown, is possible only because *patterns* of light and dark alternations measured in terms of spatial frequency (temporal frequency for auditory signals) are encoded in resonance patterns in the cortex. Cells themselves are complexes of vibrating energy that resonate within very predictable ranges to other frequencies. Most importantly, because the frequency patterns of the various sub-systems, including even a single cell in the visual cortex, conform to the predictions of Fourier analysis, then the brain must store information in the way holograms do.

A similar type of modeling is developed by Dana Zohar in her recent work, *The Quantum Self*, which not coincidentally carries a favorable review by David Bohm.

> I suggest that the electrical firing that constantly takes place across neuron boundaries whenever the brain is stimulated might be providing the energy required to jiggle molecules in the cell walls, causing them to emit photons. By way of such signals, the molecules in any given cell walls, and in thousands of nearby ones, could communicate with each other in a 'dance' that begins to synchronize their jiggling (or their photon emissions). At a critical frequency they would all jiggle as one, going into a Bose-Einstein condensed phase. The many dancers would become one dancer, possessing one identity.
>
> At that crucial point . . . the movements of the synchronized molecules within neuron cell walls (or photons emitted by them) would take on quantum mechanical properties--uniformity, frictionlessness, (and hence persistence in time), and unbroken wholeness. In this manner they would

generate a unified field of the sort required to produce the ground state of consciousness. The phase shift, then, is the moment when 'an experience' is born.[15]

Some of the best experimental evidence for, as well as an excellent introduction to, a holographic model is found in biologist Paul Pietsch's book, *Sufflebrain*.[16] In a series of critical experiments, Pietsch set out to disprove a holographic model in favor of a structuralist model, but concluded by rejecting the latter. His basic experimental format consisted in removing, reversing, scrambling and even transferring to frogs sections of salamander brains in order to determine the conditions under which feeding patterns would remain intact. Every structuralist prediction failed, and every 'hologramic' (his term of preference) prediction was confirmed. Function survived the severe alteration of structure.

In a provocatively titled essay 'Is Your Brain Really Necessary?' neurologist John Lorber surveyed the results of brain destruction caused by hydrocephalia.[17] In one case the visual cortex was destroyed, yet the patient retained normal vision. In another only 1/45 of the entire cortex remained intact, yet the patient was normal in every respect save intelligence, which was unusually high. Such a case not only lends support to something like a holographic model, but also has the makings of a suggestive regress. If 1/45 is sufficient for normal thinking, imaging, feeling, and recall, why not 1/60 or 1/90 or even a portion barely visible?

Norman Malcolm has argued that any explanation of memory retention in terms of neural coding presupposes psychophysical isomorphism between the initial experience and its physical 'trace' or between its trace and a current recollection.[18] However, because such isomorphism is doomed to a fundamental incoherence, no account of neural memory storage can succeed. This is not the place to examine Malcolm's reasons. However, it is worth pointing out that, even if they are essentially sound, their applicability to Pribram's model (which he does not examine) appears to be quite limited. A brief explanation of why this is so will serve the limited purposes of rescuing Pribram's proposal from an *a priori* objection and of clarifying the model itself.

To begin, Malcolm assumes that the concept of a trace at issue is that of a localizable neurochemical entity. Yet Pribram rejects this conception in favor of a far more complex model of nonlocalized coding. Put differently, Malcolm assumes that structural isomorphism is at issue, whereas Pribram rejects structural isomorphism in favor of patterns exhibiting common mathematical form. Malcolm also objects to an assumption that permeates the language of most neuropsychologists, namely, that neural traces *cause* distinctive mental representations. However, Pribram avoids this move in affirming that mental representations *are* interference patterns.

Perhaps Malcolm's strongest argument is that the concept of a single fixed trace is incapable of accounting for the varied representational aspects of memory that are tied to context-dependent differences of interest, perspective, and ability. A simple case. I perceive the (reversible) stairs as 'going up.' Upon thinking about it, I later recall them as 'going down.' Yet, neurologically speaking, the hypothetical trace is one and the same entity. How is such variance possible? Interestingly, Pribram is motivated by similar concerns one of which has the reverse structure of the example just cited. He asks, for example: 'How is perceptual constancy possible when the neural input deriving from a moving object is continuously changing?' Moreover, lack of variability in standard hard-wired brain models is partly what drove him to a holographic model in the first place. Such modeling, he claims, permits virtually limitless variability of mental representation under identical or very similar stimulus conditions. The extent to which he can make good on this claim remains to be seen. Malcolm's *a priori* objection, however, need not undermine the venture from the outset.

One of the implications of Pribram's work is that the already considerable body of evidence indicative of equipotentiality and/or neural redundancy that leads to a holographic model also forces a revision in materialist accounts of mental states and events that are in principle tied to the success of localization strategies. To be so tied is to hold out the prospect of discovering the sufficient *and* necessary neural correlates of those entities nominally designated as 'mental,' e.g., thought, feelings, sensations, etc. But if the evidence for a holographic model is taken seriously, we can predict that those correlates deemed sufficient will not, strictly speaking, be necessary. They will not be necessary if, and to the extent that, neural excitation *in a different place* in principle can produce the same effects. In the case of memory our recollections are no more one place than another, and this is beginning to appear to be the case with other classes of mental entities, such as perceptual images. Localization is a useful point of departure in brain science, but an increasingly elusive criterion upon which to stake a materialist metaphysics.

The dualist, of course, has long urged that experiences such as depression, do not admit of predicates ascribing spatial location. The virtue of the holographic model is that it shows how the nonlocalizability of mental states, in a limited sense, is both possible and reasonable without conceding anything to a dualist ontology.

The holographic model might best be described as a type of 'energy monism.' Pribram adopts a field theory of matter which stipulates that mass is a function of highly compressed and intense patterns of energy. And while he does not state matters thusly, in effect he proposes a reduction of *both* states of consciousness and brain states to an underlying locus of vibrating energy and interference patterns. The difference between the experience of pain and the activated C-fiber which occasions it is therefore seen as one of degree, not of fundamentally distinct kinds. Simply put, some patterns of energy with distinctive frequencies, amplitudes, and interference features give rise to material particles while others occasion states of consciousness, just as, for example, sound, color, and x-rays are occasioned by different frequencies. Pribram would argue that standard reductive materialism ('The state of depression is strictly identical with activated C-fibers') is false for essentially the same reason that colors are not sounds (or x-rays, etc.). But this does not entail dualism, since the depression/C-fiber distinction is perhaps no greater than the color/x-ray distinction, and the latter difference is grounded in a single underlying spectrum of electromagnetic energy.[19]

If depression is not identical with firing C-fibers, and dualism is not a viable alternative, then why does C-fiber stimulation occasion depression? While admittedly speculative, I think Pribram's most plausible answer would be to stipulate that the C-fiber stimulation sets off distinctive resonance patterns of a much higher (lower?) order with which it shares a common mathematical form as, for example, striking low C sets up resonances in only certain other strings. Those interference patterns correlated with states of consciousness may involve frequency ranges not currently recognized by physics, as suggested by the concept of 'life fields' or of 'bioplasmic' energy. In Pribram's view, however, it would be some form of energy that in principle falls within the domain of physics.

In defending holographic modeling against the kind of objections raised by Malcolm and others of a Wittgensteinian persuasion, it should be stressed that I am not arguing for a *reduction* of consciousness to the wave-forms and frequencies of a physical or even of quasi physical energy. I am not suggesting, for example, that thoughts (which cannot be extensionally divided) or depressions (which do not admit of predicates of spatial location) are fundamentally a 'bunch of vibrations.' Many features of consciousness, such as intentionality and the privacy of experience, resist such reduction.

I would argue instead for a double-aspect theory in accordance with which consciousness is viewed as the 'inside' of energy and energy as the 'outside' of consciousness.[20] For example, emotions may be said to become manifest to an external observer in an energetic fashion at any of three different levels: (1) the phase entanglements and holographic interference patterns of correlated neuron firings and photon exchanges accessible to the aided eye of the neurobiologist; (2) overt action stressed in behaviorism, or; (3) possibly an auric field perceived by a clairvoyant.

However, this is not to say that one's emotions, as experienced from the inside, are nothing more than phase entanglements, behavior, or auric fields each of which is available in principle to an outside observer. It is to suggest that, insofar as the elements of consciousness are correlated with something of a physical or quasi-physical sort, that 'something' (per Pribram, *et. al.*) is more of an energetic or vibrational nature than the localizationist assumptions of standard neuroscience seem to allow.

Morphogenetic fields in biology

Biological morphogenesis (the emergence of characteristic and specific *form* in living organisms) involves three related issues pivotal in the debate between vitalists, mechanists, and organicists. The first is epigenesis, the fact that new and more complex structures appear which cannot be explained in terms of the unfolding of simple structures present in the fertilized egg at the beginning of development. The second is regulation, the fact that if part of a certain system is removed the system continues development toward a more or less normal structure (e.g., having a young sea-urchin embryo yields a smaller but complete adult). The third is regeneration, the fact that many adult organisms are able to replace damaged or destroyed parts (e.g., a newt will replace its surgically removed lens).

Especially since the discovery of the double-helix, most biologists have held out the belief that the key to morphogenesis is coded in DNA molecules. However, in light of thirty years of intensive investigation it is beginning to appear that the kind of information DNA molecules are capable of yielding has built-in limitations for explaining morphogenesis. For example, virtually the same code is known to yield both chimpanzees and humans, and in fact greater gene differences are evident between more closely related species of mice or fruitflies.[21] Moreover, since identical DNA structures are distributed throughout our bodies, how is it that some give rise to arms, others to legs, kidneys, and so forth in patterns requiring pinpoint timing? When the code is the same, what could account for differentiation?

Research devoted to identifying the exact physico-chemical triggers for the complex and diverse patterns of differentiation in an organism thus far has yielded many proposals but no general answers. And even if we assume that such mechanisms will be identified, there appears to be no way in principle to account for their distinctive patterns and triggering characteristics except by a circular appeal to DNA. For example, in some plants the hormone auxin plays a key role in vascular differentiation. Yet the protein breakdown involved in such differentiation appears to play the central role in producing and distributing auxin.

Despite the tremendous advances in our understanding and control of genetic coding, the metaphor of a complete 'blueprint' suggests that the DNA has been endowed with more explanatory power (as elusive perhaps as the engram) than it is capable of exhibiting. Subject to variation of detail, we know only that it provides a sequence of chemical letters which are translated via RNA into distinctive sequences of amino acids in proteins. Part of the DNA controls the structure and part the actual synthesis of the proteins necessary, say, for a human body. But nothing we know of explains why cells organize the way they do in embryogenesis, much less in regulation and regeneration. To exploit a metaphor, we might say that it produces the ingredients for a dwelling (steel, lumber, etc.) in the appropriate quantities, but it tells us nothing of the final form, a fifteen-room mansion or a two-room hut. Different phyla indeed exhibit distinct chemical sequences, but this is to say no more than that coding is necessary but not sufficient for explaining morphogenesis.

In addition to morphogenesis proper, there are numerous behavioral anomalies encountered in biological and related contexts. For example, in 1952 Japanese ethologists distributed dirty sweet potatoes to a monkey troop (*macaca fuscata*) on a small Pacific island. The unwashed potatoes were first rejected until by chance a young female dipped one in a stream. Over a period of 6 years others learned this behavior until a critical mass of approximately 100 monkeys were washing their potatoes, at which point within a twenty-four hour period the remaining members suddenly took up the behavior. It so happened, however, that monkeys of the same species on a neighboring island subsequently given dirty potatoes reportedly almost immediately replicated the behavior of their cousins on the first island without the necessity of learning.

Was this chance or was information in some attenuated sense transmitted? It remains to be seen if there are any legitimate '100th Monkey Phenomena,' since it now appears that the original scenario was incorrectly reported.[22] However, even as a hypothetical example, this rapid spread of information nicely illustrates the *kind* of anomaly for which some type of field modeling is called.

That geneticists may have issued too many promissory notes regarding morphogenesis or that there are plenty of unresolved puzzles regarding animal behavior are not, of course, novel themes. They are, however, developed in a forceful and updated fashion by biochemist Rupert Sheldrake in *A New Science of Life*, which introduces a novel teleological model termed the 'hypothesis of formative causation.'[23]

Morphogenesis, he argues, requires us to posit appropriate *fields* which complement genetic codes by guiding the development of biological form. Sheldrake's defense is not an outdated attempt to resurrect hylomorphism. It should be of interest to both biologists and philosophers that he develops a detailed model of how such fields are causally efficacious, spells out a variety of test implications, and demonstrates knowledge of the philosophical positions that historically and logically bear most directly upon his thesis and its relevance for neo-Darwinian thought.

A summary statement of his hypothesis includes the following key tenets. There are as many fields as there are species of organism, as well as fields for inorganic elements and compounds. Such fields affect learning and motor activity as well as physical form. They exist 'outside' of space-time, do not transmit energy, and interact with local fields involving known physical forces.

Their causal influence derives in part from their distinctive 'resonance patterns' which require physico-chemical structures to conform to certain spatial forms as, for example, an electromagnetic field (which does not transmit energy) requires iron filings to conform to a pattern. The fields are derived from and reinforced and modified by the objects they influence. They are neither fixed not eternal.

Sheldrake's hypothesis is purportedly testable. But what type of empirical evidence in principle could suggest morphogenetic fields, so conceived? For example, there is a considerable body of evidence for the existence of electrodynamic fields surrounding living organisms, especially plants.[24] Yet Sheldrake cuts short this line of comparison: 'Electromagnetic fields . . . depend upon the *actual* state of the system--on the distribution and movement of charged particles--whereas morphogenetic fields correspond to the *potential* state of a developing system and are already present before it takes up its final form' (p. 77). This passage invites comparison with Aristotle's formal cause or with Driesch's entelechy. Yet the fields' origin, mutability, and non-local status all belie such comparison.

Sheldrake's more indirect line of evidence derives from the stipulation that the causal force and qualitative nature of the fields reflect the quality and range of the objects they influence. The fields are subject to change and even dissipation over long periods of time. They reflect environmental impact upon species, chance mutations, and the phasing out of species. The more members of a class that exist, the greater is the field force and the more likely that physical forms or behavior of that type will continue to appear. Some fields are so well established, say, for hydrogen atoms or pine trees, that they appear to function as immutable laws of nature. Yet Sheldrake contends that their influence must be understood probabilistically, a wise move in light of the existence of two-headed cows and armless babies.

From this conception of a field, he derives the test implication that after the introduction of a comparatively novel physical form or type of behavior, and given minimally necessary conditions of proliferation, we should be able to observe replication in times and places that defy normal spatio-temporal barriers or chance expectations. Sheldrake both points to instances where this has happened and suggests numerous ways it can be put to the test.

Chemists long have been aware, for example, that after the spontaneous synthesis of a novel crystal there is a tendency for identical forms to appear in comparable solutions in sealed jars in other wings of a building at rates that defy chance parameters. Their typical explanation is that contamination has occurred in which some of the initial crystals unknowingly were transmitted to other solutions. By contrast, Sheldrake interprets this as the result of the increased influence in the field which operates non-locally over normal spatial barriers. More importantly, however, he spells out in detail the experimental conditions necessary in testing his hypothesis to rule out the contamination explanation (pp. 104-108).

Another type of test implication is that a specific task should be more readily learned by each successive generation of a carefully inbred species. Lamarckianism of course, also predicts this result as a consequence of the inheritability of acquired characteristics, and has been put to the test in a variety of formats over a fifty-year period. In a definitive twenty-year study, W.E. Agar confirmed earlier findings (successive generations of an inbred species of rat did learn a specific task more readily), but rejected Lamarckian interpretations because the *untrained* family members of successive control groups showed virtually identical increases in the rate of demonstrating performance of that task.[25]

Since the purpose was to put Lamarckianism to a critical test, the resulting anomaly has received little attention. Yet the results are precisely what Sheldrake's hypothesis would predict, i.e., that the emerging field impartially affects both trained *and* untrained rats. It thus differs from Lamarckian and Neo-Darwinian models.

Sheldrake argues that a materialistic interpretation of the fields, although still possible, is not preferable. I would submit, however, that even granting his arguments for some type of field, modified materialism is still the preferred interpretation. To begin, if the kind of evidence for the existence of morphogenetic fields is empirical, then our description of the nature of the fields in principle ought to be subsumable within the domain of science. He would respond that since their causal influences are operative in ways that defy normal spatial barriers it is reasonable to describe the fields as 'outside' space, i.e., outside the scope of physics. Yet the evidence he cites actually suggests the existence of initially small but expanding local fields. For example, the accelerated proliferation of a novel crystal structure occurs in the same general area of the initial spontaneous formation, not in another country. The same is true for anomalous increases in rates of learning that he cites.

Moreover, if the fields are linked to known energy systems and could not exist without that two-way interaction, then the interests of scientific progress and conceptual clarity will be served better by assuming that presently unknown types of field are involved, rather than that the nonphysical is acting upon the physical. Indeed, his suggestive model of interaction based upon conformity to frequency or resonance patterns is itself borrowed from our understanding of how physical energy manifests itself, i.e., in cycles per second.

It is perhaps too easy to criticize Sheldrake on philosophical grounds given the magnitude of the problem he addresses. His significance rests not so much with the admittedly tentative answers he proposes as with the kinds of issues he examines and the general type of modeling that may be necessary to resolve them. He does show not only that causal influences may work in ways that we have not suspected, but also that they may require some radical conceptual revision. In conjunction with developments discussed earlier, his work exhibits a move toward an expanded materialism in which fields, forces, and frequencies progressively assume roles formerly given to the position, configuration, and movement of units of matter (in this case genes).

Near death experiences in medicine/psychology

It is by now a familiar story. The patient undergoes a cardiac arrest, lapses into a state of unconsciousness, and approaches a condition of clinical death, but after a few minutes is resuscitated. During this time she experiences herself floating out of her body, observing or hearing various goings-on in the operating room, moving through a gray mist or dark tunnel toward a white light, possibly encountering deceased friends or 'spirit guides' in a humanlike form, and enjoying peacefulness and major attitudinal changes that carry over into daily life. Subject to some variation this is the classic near-death experience (NDE).

Since the publication of Raymond Moody's *Life After Life* in 1976 the interest in and literature about NDEs has grown enormously. This is evidenced by a major national symposia (Psychology and Psychiatry), articles in prestigious scientific journals (both sympathetic and critical), a new interdisciplinary *Journal of Near-Death Studies,* curricular inroads at major medical schools, and one of the most extensive Gallup polls ever undertaken which indicated that NDEs are far more common than anyone had suspected.[26] The area is rife with opportunities for philosophical involvement, but to date most of the philosophizing is being done by nonphilosophers.

What *is* a near-death experience, that is, which features are logically necessary and/or sufficient for NDEs? How sound are the arguments supporting objectivist interpretations (persons actually leave their bodies) and subjectivist interpretations (NDEs are nothing but an interesting form of projective hallucination)? What, if any, are the implications of NDEs for the mind-body problem and survival of bodily death?

Many philosophers no doubt share the widespread medical opinion that NDEs are merely a neuropsychiatric phenomenon. Why examine the arguments of a non-issue? To the contrary, there is by now sufficient evidence to justify more than a cavalier dismissal of NDEs on the grounds that, after all, the brain is capable of playing many poorly understood tricks.

The charge that the religiously inclined naturally will encounter the 'next life' overlooks the fact that as many lifetime atheists and materialists have NDEs. That near-death experiencers achieve ego-gratification through some public acknowledgement does not take into account that the great majority of those interviewed (by now several thousand) fear ridicule and demand anonymity.[27]

Moreover, the least defensible aspect of the experience, e.g., that Jones claims to have encountered, say, the Virgin Mary, tends to be emphasized while it is overlooked that Jones also may have made verifiable claims about events in the emergency room. That 'moving through a dark tunnel' is merely reliving 'travel down the birth canal' (Carl Sagan's observation) not only does not explain other key features of the NDE but also overlooks the fact that a sizable number of near-death experiencers were born by Caesarian section.[28]

By far the most sustained critique is psychopharmacologist Ronald Siegel's essay 'The Psychology of Life After Death' which contains many interesting facts about the physical genesis of various types of hallucination together with rather standard observations regarding the origin of beliefs about survival.[29] It does not, however, address much of the evidence accumulated thus far and assumes throughout that the only issue is that of discovering the causes of such hallucinations.

The most sustained case of for an objectivist interpretation is physician Michael Sabom's book, *Recollections of Death: A Medical Perspective*.[30] Sabom, a converted skeptic, does not claim proof. He does, however, marshal evidence that deserves serious scrutiny. As important as the evidence per se, however, is the fact that he has defined much of the territory over which objectivist vs. subjectivist interpretations will be fought. He confirms earlier findings that NDEs do not correlate significantly with any sociological or psychological variables. I draw attention, however, to his more distinctive contributions.

Sabom's first line of argument involves the verification of claims made by patients that they saw or heard certain events when they were supposed to have been unconscious and close to death (pp. 80-116). He goes to considerable lengths, for example, to show that certain observation claims could have been made only from a vantage point distinct from one's body. Interviews with operating room personnel and reviews of available medical records (time of arrest, etc.) obviously play a critical role here. In some cases the quality and quantity of descriptive detail is remarkable, e.g., a surgeon's wearing mis-matched socks. Against such detail, Sabom asks: 'How could such information be acquired?' He considers alternative explanations in some detail, e.g., 'Is it likely that an itinerant farmer could provide a medically accurate description of triple by-pass heart surgery from having watched *General Hospital*?'

To be sure, the evidence of this type is far from overwhelming, and Sabom himself views it as merely exploratory and suggestive. Suppose, however, that it continues to be developed in considerable quality and quantity of detail, including, for example, precise readings from automatically recorded instrument gauges or the congenitally blind seeing for the first time. If such proves to be the case, then hallucination hypothesis would have to give way to a conceptually revolutionary objectivist interpretation.

In a second line of argument Sabom takes proposed medical explanations to task (pp. 151-178). In what essentially is an application of Mill's canons, Sabom demonstrates that each of a dozen proposed explanations, e.g., temporal lobe seizure, hypoxia, endorphin release, etc., not only fails to account for most features of the NDE but also should be correlated with symptoms that in fact are absent consistently in NDEs. Of course, he is the first to point out that some neurophysiological trigger *might* be discovered. He argues merely that, so far, we are not even in the ball park.

He also should have pointed out that there are certain logical limitations attending such a discovery. The NDE per se does not become hallucinatory merely by virtue of being correlated with a neurochemical variable. How would such a variable explain the acquisition of information described above?

The third and potentially strongest line of argument, which Sabom does not develop in any detail, involves the correlation of NDEs with flat brain-wave activity. To date, there are few, if any, reliable cases on record due to the fact that emergency and operating room procedures seldom require the use of EEG machines. However, if such correlations are forthcoming, then the critic will be hard-pressed to maintain that the patient was hallucinating by means of a brain that was electrically 'dead.' For any type of hallucination should result in some brainwave activity.

The methodological and conceptual issues posed by NDE research in some respects are analogous to those found in parapsychology. Indeed, parapsychologists have been aware of NDE research from its inception. There is some cross-fertilization. However, the key issue raised by largely descriptive NDE research, namely, the possibility of out-of-body experience, typically is confronted experimentally in psychical research, independently of the medical crisis atmosphere of NDEs which thus far has attracted mostly physicians and psychologists with little background in parapsychology.

Since NDE and OBE research intersect, a brief up-date on the latter is in order. Philosophers who argue that any form of OBE (survival, witnessing one's funeral, etc.) is logically impossible might wish to take into account Karlis Osis' highly suggestive OBE research.[31] Briefly, Osis has had at least one subject who not only scored impressively when asked to 'project' to another room and describe randomly selected pictures (in one series 114 correct calls out of 127 possible), but also who, within narrowly prescribed time frames, both intentionally and unintentionally triggered devices which measure various physical forces, e.g., strain gauges and magnetometers. Such experiments, if successfully replicated, will render OBE hypotheses (and, by implication, those associated with near-death experience) more palatable, *a priori* arguments to the contrary.

Resistance to objectivist explanations of NDEs (and OBEs in general) on philosophical grounds stems from arguments against survival or arguments against dualism or both. However, it may be overlooked that, from a logical point of view, the prospective evidence that suggests short-term survival, viz., seeing one's body at a distance, actually *counts against dualism*. For to see at all, whether with physical or 'ethereal' eyes, is to see from a point in space. The dualist's mind, however, is by definition non-spatial, i.e., it does not occupy a point or localizable volume in space. Therefore, whatever leaves the body cannot be non-physical. Indeed, since the entity involved in principle must be localizable--it needn't be directly observable--then *some* type of energy is strongly suggested. This general conclusion is reinforced by Osis' detection of local forces at work away from the subject's body. Modified materialism here clearly seems preferable to dualism. At the least, however, NDEs should not be rejected in the mistaken belief that out-of-body experiences entail dualism.

Concluding remarks

Some concluding remarks are in order regarding the general significance of the issues examined in this essay. To begin, there is a cluster of concerns with the conditions underlying the acquisition, retention, or transfer of 'information.' In scientific contexts the term is used loosely, sometimes to mean propositional knowledge and sometimes to mean 'information carrier' in the sense that the properties of certain physical objects or processes, such as electron flow or genes, are arranged so as to permit translation into a cognitive domain. The concerns are expressed by the following questions: How could a person acquire information while in a state of unconsciousness or temporary electrical brain death (Sabom)? How could it be stored equipotentially (Pribram)? How could it be transferred in ways that violate normal spatio-temporal barriers (Sheldrake, EPR-effect)?

A second and closely related common thread is that, in the contexts described, the ideal of *localization* in different ways is called into question. In each case what occurs in one place carries an intrinsic reference to events in another place in ways that violate normal explanatory models. For example, we saw that the spin-state of one electron is dependent instantaneously upon changes in its formerly paired partner, in principle miles or even light-years away. And a single recollection, such as of my slipping on the rug, does not appear to be located in one place in my brain, but is instead spread around in different places. Further, the fact that many cells with an identical genetic code form different structures at different times and places may depend upon other factors outside the local configuration of chemical letters. (Some of Sheldrake's concerns are simply larger versions of Pribram's concern with equipotentiality. Compare 'How can organs regenerate'? with 'How can memories be retained when their presumed physical basis is destroyed?')

Finally, judgements of unconsciousness or even temporary electrical brain death may be called into question by the possibility that the owner in some attenuated sense was at that time 'gathering information elsewhere.' How can a person be contingently identical with the body there, if he observes it from a point elsewhere?

Attending the anomalies of localization in each case is the failure to provide an acceptable *mechanistic* explanation of the phenomena involved. Of course, such explanations may be found, and the prospects vary with each case. On the whole, however, they do not look promising. This is not to say that causality has broken down, i.e., that there are spontaneous, uncaused events in the universe. Rather, it is to suggest that it works in more interesting ways than are captured in models that stress the position, configuration, and physical contact of basic physical units, be they electrons, genes, or hypothetical engrams. The failure of mechanistic models in these areas occasions the need for stressing the kind of interconnectedness suggestive of an organic philosophy of nature which adopts a stance between the extremes of reducing objects to their relations (the doctrine of internal relations) or of assuming them to be completely comprehensible in abstraction from their relations (atomism). Certainly the emphasis we have seen given to fields, frequencies, and interference patterns incorporates, albeit in different ways, a greater commitment to interconnectedness.

This brings us to a final observation, namely, that any failure of mechanistic explanation in the cases we have described does not entail the failure of a materialist ontology. Idealism or dualism may be suggested by some of the phenomena, but neither possesses any decisive advantages over an expanded materialism (or perhaps a double-aspect theory) in which fields and frequency patterns play progressively greater explanatory roles. Some of the 'energies' involved may be reducible to one of the four primary forces in nature, and some may require an expansion of our concept thereof. But this is the stuff of interesting science, not a fundamental shift in ontology. Attending the prospects for an expanded materialism is a discernible tendency to view interaction between parts as dependent on the state of the whole they comprise (at whatever level of abstraction we choose), material stuff as the expression of energy, and physical discreteness as continuous with underlying non-local fields. Such proposed distinctions of degree are more reminiscent of Lovejoy's Great Chain of Being than of any sharp distinction between appearance and reality.

It remains to be seen whether the developments examined in this essay constitute key elements in an emerging paradigm shift, if I may be permitted one unexamined use of the expression. Bohm, Pribram, and Sheldrake, among others such as Fritjof Capra, view their work as pointing toward such a shift.[32] Whatever the outcome, it cannot fail to be enhanced by more extensive philosophical scrutiny.

Notes

*I thank James Humber, Robert Arrington, William Bechtel, and Robert Almeder for helpful suggestions and criticisms.

1. Albert Einstein, Boris Podolsky, and Nathan Rosen, 'Can Quantum-Mechanical Description of Physical Reality be Considered Complete?' *Physical Review*, vol. 47 (1935), pp. 777-780.

2. J.S. Bell, 'On the Einstein Podolsky Rosen Paradox,' *Physics*, vol. 1 (1964), pp. 195-204.

3. For a review of the relevant experimental literature, including the critical '72 test by John Clauser and Stuart Freedman using photomultiplier effects, see B. d'Espagnat, 'The Quantum Theory and Reality,' *Scientific American* (December, 1979), pp. 158-181. See also David Mermin's review, especially of the more recent work by Alain Aspect, in 'Is the moon there when nobody looks? Reality and the quantum theory,' *Physics Today*, vol. 38 (1985), pp. 38-47.

4. Why such attempts are unsuccessful is demonstrated by Clifford Hooker in 'Concerning Einstein's, Podolsky's, and Rosen's Objection to Quantum Theory,' *American Journal of Physics*, vol. 38 (1970), pp. 851-857. I also have heard it maintained by competent physicists that, by analogy, the physical interpretation of the relevant mathematics is no more significant than the fact that, if two balls, one black and the other white, were placed in random, unobservable trajectories, the eventual interception of one would enable us to infer the color of the other. Why, then, could we not change the color of one merely by changing the color of the other?

5. John Clauser and Abner Shimony, 'Bell's Theorem: Experimental Tests and Implications,' *Reports on Progress in Physics*, vol. 41 (1978), p. 1881.

6. Clauser and Shimony, *Ibid.*, see Berkeleyian idealism as the logical alternative as does Eugene Wigner in *Symmetries and Reflections* (Bloomington: Indiana University Press, 1975). For evidence against long-distance psychokenetic action, which is more directly relevant to any proposed connection between idealism and the EPR effect, see Joseph Hall, Christopher Kim, Brain McElroy, and Abner Shimony, 'Wave-Packet Reduction as a Medium of Communication,' *Foundations of Physics*, vol. 7 (1977), pp. 759-767.

7 See for example, Henry Stapp, 'Are Superluminal Connections Necessary?' *Il Nuovo Cimento*, vol. 40B (1977), pp. 191-205. Physicist Nick Herbert argues in the affirmative in *Quantum Reality* (New York: Anchor, 1987).

8 J.L. Mackie, *The Cement of the Universe* (Oxford, 1974), p. 19. For a useful though somewhat dated review of the literature, see Bob Brier, *Precognition and the Philosophy of Science* (New York, 1974). The challenge that the EPR effect poses

for causal explanation is discussed by Wesley Salmon, Bas van Fraassen, and Philip Kitcher at the December 1985 APA symposium and included in *The Journal of Philosophy*, vs. LXXXII (1985), pp. 632-654.

9. Henry Stapp, 'S-Matrix Interpretation of Quantum Theory,' *Physical Review*, vol. D# (1971), pp. 1303-1320.

10. William S. Condon, 'Multiple Response to Sound in Dysfunctional Children,' *Journal of Autism and Childhood Schizophrenia*, vol. 5 (1975), p. 43. Whether we interpret this as simultaneous causation or as the effect preceding the cause depends, of course, upon the exact time we assign to the occurrence of the cause. There are methodological and conceptual issues in so doing (When does speech begin?), but they do not affect Condon's observations of listener reactions.

11. David Bohm, *Wholeness and the Implicate Order* (London, 1980), p. 175. B. d'Espagnat, *op. cit.*, similarly concludes that the violation of Einstein's assumptions (regarding locality) seem to imply that in some sense all particles 'constitute an indivisible whole' (p. 180).

12. Karl Pribram, 'The Neurophysiology of Remembering,' *Scientific American*, vol. 220 (1969), pp. 73-86.

13. Much of this evidence is surveyed in Pribram's key work, *Languages of the Brain* (Englewood Cliffs: Prentice-Hall, 1971).

14. F. Bartlett and E.R. John, 'Equipotentiality Quantified: The Anatomical Distribution of the Engram,' *Science*, vol. 181 (1973), pp. 764-767.

15. Danah Zohar, *The Quantum Self* (New York: William Morrow, 1990), p. 85. Pribram's arguments for holographic modeling in perception in particular are contained in his 'Holonomy and Structure in the Organization of Perception,' in U.M. Nicholas, ed., *Images, Perception, and Knowledge* (Dordrecht: Reidel, 1977). Also see his overview in 'Holographic Memory,' *Psychology Today* (February, 1979). Pribram does not attempt to overturn or by-pass localization strategies in discovering psychophysical correlations. His thesis is, rather, that a complete explanation of information processing requires reference to a deeper holographic structure. His demonstration that apparent cellular reaction in the perception of geometric form is itself a function of frequency decoding illustrates the complementary nature of his modeling strategy.

16. Paul Pietsch, *Sufflebrain* (Boston, 1981), especially chapters I-IV.

17. John Lorber, 'Is Your Brain Really Necessary?' *Science*, vol. 210 (1980), pp. 1232-1234. The evidence from brain scans is not conclusive, but it is strong enough to have stimulated major discussion and the search for other explanations.

18. Norman Malcolm, *Mind and Memory* (Ithaca: Cornell University Press, 1977). Subsequent page numbers indicate where, from among many places in Part II, discussion of the relevant issue can be found. Stephen Braude critiques Kenneth Ring's holographic modeling of Near-Death Experiences (see Section IV) along similar Wittgensteinian lines in his 'Holographic Analysis of Near-Death

Experiences: The Perpetuation of Some Deep Mistakes,' *Essence,* vol. 5 (1981), pp. 53-63.

19. In making this move, he would have to show that the pain/energy identification in principle carries more manageable problems than those typically encountered in pain/C-fiber identifications, perhaps by showing that the emergence of certain raw-feel properties from interference patterns, conceptually speaking, is no more problematic than the emergence of certain material properties from underlying exchanges or compressions of energy.

20. This proposal is developed in more detail in my 'Beyond Dualism and Materialism: A New Model of Survival,' in *What Survives*? ed. by Gary Doore (Los Angeles: Jeremy Tarcher, Inc. 1990).

21. Accordingly, a Leibnizian might inquire why we are human and not ape. See M.C. King and A.C. Wilson, 'Evolution at Two Levels in Humans and Chimpanzees,' *Science,* vol. 188 (1975), pp. 107-116.

22. An informal exposition of this phenomenon is given by biologist Lyall Watson in *Lifetide* (New York, 1979), pp. 146-148 and critiqued by Ron Amundson in 'The Hundredth Monkey Phenomenon,' *Skeptical Inquirer,* vol. IX, (1985), pp. 348-356.

23. Rupert Sheldrake, *A New Science of Life* (Boston: Houghton Mifflin Company, 1982). Sheldrake is a Cambridge trained botanist and biochemist and spent a year at Harvard as a Frank Knox Fellow studying philosophy and the history of science. *The New Scientist* described it as 'an important scientific inquiry into the nature of biological and physical reality,' while *Nature* described it as the 'best candidate for burning there has been for many years.' Sheldrake refines his hypothesis in *The Presence of the Past* (New York: Random House, 1988). See also Susan Blackmore's critical review in the *Parapsychology Review,* vol. 16 (1985), pp. 6-10 and Stephen Braude's critique in 'Radical Provincialism in the Life Sciences,' *The Journal of the American Society for Psychical Research*, vol. 77 (January, 1983). Experimental support for the thesis is described in 'Formative Causation: The Hypothesis Supported,' *New Scientist* (October, 1983).

24. An informal review of the evidence and theoretical implications is given by Yale biologist Harold Burr in *Fields of Life* (New York, 1972). See also the older but still theoretically relevant article co-authored with F.S.C. Northrup 'The Electrodynamic Theory of Life,' *Quarterly Review of Biology* vol. 10, 1935, pp. 322-333.

25. W.E. Agar, 'Final Report on a Test of McDougall's Lamarckian Experiment on the Training of Rats,' *Journal of Experimental Biology,* vol. 31 (1954), pp. 307-321.

26. A spectrum of medical opinion regarding NDEs is presented in a special issue of the *Journal of Nervous and Mental Disease*, vol. 168 (1980), pp. 259-274. Yale Medical School, for example, has sponsored a symposium on the subject. Moody, a former philosophy professor turned psychiatrist, approached the topic (following Elizabeth Kubler-Ross) in a phenomenological and anecdotal fashion. Psychologist Kenneth Ring's *Life at Death* (New York, 1980) is both more informative and analytical. Gallup's findings are published in his *Adventures in Immortality* (New York, 1982).

A critique of reductive accounts is given in my 'Five Arguments Regarding the Objectivity of NDEs,' *Anabiosis*, vol. III (1983), pp. 63-77. For a model of NDEs the incorporates recent neurobiological findings, see Melvin Morse, *et. al.*, 'Near-Death Experiences: A Neurophysiological Explanatory Model,' *Journal of Near-Death Studies,* vol. 8 (Fall, 1989).

27. These and other cultural and psychological variables proposed as possible explanations are examined by Ring, *Ibid.* His was the first work to make clear the relative invariance of NDEs across major demographic categories.

28. See his *Broca's Brain* (New York, 1979), p. 304. It is examined by Carl Becker in 'The Failure of Saganomics: Why Birth Models Cannot Explain NDEs,' *Anabiosis*, vol. 2 (1982), pp. 102-110.

29. Ronald Siegel, 'The Psychology of Life After Death,' *American Psychologist,* vol. 35 (1980), pp. 911-931.

30. Michael Sabom, *Recollections of Death: A Medical Perspective* (New York, 1982). Sabom, a former cardiologist at the Emory University School of Medicine, is now in private practice. Page references are to this edition.

31. K. Osis and D. McCormick, 'Kinetic Effects at the Ostensible Location of an Out-of-Body Projection During Perceptual Testing,' *Journal of the American Society for Psychical Research,* vol. 74 (1980), pp. 319-329.

32. For an application and development of this type of modeling in medicine, see Richard Gerber's *Vibrational Medicine* (Santa Fe, NM: Bear and Company, 1988).

4 The recent case against physicalist theories of mind: A review essay

Joseph Wayne Smith

Introduction

Physicalist philosophies of mind are today the ruling orthodoxy in metaphysics. Indeed, not only do many philosophers accept that mental events are identical to physical neurological events, but some philosophers argue that mental and intentional concepts are logically and methodologically problematic and must be eliminated from scientific discourse.[1] The aim of this essay is to review the arguments of a body of literature challenging the physicalist view of the nature of mind. There are two prongs to this attack. The first prong of attack centres around the physical and conceptual difficulties in explaining how human memory is possible. The second prong of attack centres around the development of a radical critique of the correspondence thesis, the thesis that psychological phenomena correspond in a one-to-one fashion with certain brain states and processes. This attack upon materialist/physicalist accounts of mind was initiated by Henri Bergson in the early part of this century; here I have space to consider discussing only contemporary philosophers and must leave any historical inquiries for another occasion.[2]

The problem of memory: the physiological problem[3]

The most popular account of how it is possible to remember the past is the memory trace theory, and it is certainly held in one form or another by today's physicalists. A trace theory may not be the *only* possible 'mechanistic' account of memory-behaviourism is an alternative. However, most physicalists today are scientific realists and regard behaviourism as a dead end so our discussion is exclusively concerned with the trace theory of memory.

The memory trace is a neural representation of past experiences.[4] This representation is present in the brain for as long as the person remembers the experience. Therefore a trace is more than a continuous causal chain reaching from some experience to a memory response. The trace also explains the possibility of memory retention as well as storage. As Martin and Deutscher[5] point out, even if our past experiences could act directly upon us now, the concept of a memory trace still seems to be needed because without it there is no longer any reason to believe that we could remember only what we had experienced ourselves. An experience could only count as a memory experience if it was *stored* in you.

There are however a number of physical and scientific problems facing any trace theory of mind.[6] The problem of memory, as Ninian Marshall states it[7], is that in a lifetime the human memory stores 10^{10} bits of information. Storage mechanisms such as reverberating circuits, synoptic changes or changes within the protein molecules of nerve cells are inadequate for the long term storage of complex memories such as my memory of first day at school. In particular Marshall lists three major objections to trace theories of memory, objections which he regards as stumbling blocks for conventional theories of memory:

(i) There are not enough synapses to store all memories. And if smaller objects are suggested, such as protein molecules, then if their constitution can be changed by neutral action it is unlikely to be stable against metabolism and random 'noise'.

(ii) A complex thought or experience involves the activity of millions of neurones. Any memory trace left would involve as many neurones. And each neurone must carry traces, from thousands of millions of memories. To recall a memory would involve the simultaneous selection of each bit of its trace from the appropriate neurone. This vast feat of co-ordination has no conceivable mechanism in current physics.

(iii) Lashley's Law of Mass Action. Cortical lesions cause a memory loss whose size depends on the size of the lesions but not on its position. The memory loss is greatest for complex memories. What is more, after a time the memories may return almost completely. So the same neurones are not always involved in the recall of a given memory. No wonder that Lashley remarked, 'I sometimes feel, in reviewing the evidence on the localization of the memory trace, that the necessary conclusion is that learning just is not possible'.[8]

John Beloff[9] identifies three general classes of evidence that have been taken to support trace theory. They are: (1) the study of memory in brain damaged patients; (2) direct brain stimulation to specific areas of the brain by electrodes and (3) computer simulation. In the first case Beloff is sceptical of claims that Wilder Penfield has demonstrated the existence of traces.[10] Penfield found that the stimulation of a particular spot on the temporal lobe of a patient awaiting brain surgery, meant that the patient would remember vividly some long forgotten past event. These would seem to be a forceful empirical demonstration of the truth of trace theory if the same event was remembered when the same area of the brain was stimulated again. However Penfield himself reported that when the same spot was stimulated a few minutes later, the memory recalled by the patient was quite different and other researchers found that the reported memory depended upon the patient's thoughts at the moment of stimulation. Hence no one-to-one correspondence between a specific brain trace and a specific conscious experience has been demonstrated.[11]

Studies of cerebral lesions may appear to offer some empirical evidence for the existence of memory traces. Bergson responded to this objection in his book *Matter and Memory*. In his discussion of aphasia he argued that the lesion disturbs the whole sensory-motor complex.[12] It is the disturbance of the holistic integrity of the brain which results in the loss of the person's cognitive ability. As Braude has pointed out, if there are no specific internal mechanisms to explain human cognitive abilities, it does not follow that occurrent or dispositional mental states cannot be altered by physical or chemical changes.[13]

Beloff also points out that artificial intelligence and computer simulation of memory does not add any support to the thesis that human memory involves memory traces, indeed it is to beg the question to say that the trace theory of memory must be true because computers use physical traces. This is the case even if computers are regarded as intelligent and having a 'mind'. It is also the case even if computers were devised which did not depend upon information inputs and retrieval being in a specific form. Furthermore, human memory is problematic because a given experience can be encoded or decoded in an indefinite number of ways with almost anything constituting a manifestation of memory. The significance of this fact has been recognised by the philosophical critics of trace theory, to which we now turn.

The philosophical critique of trace theory[14]

Roger Squires[15] and Normal Malcolm[16] are two philosophers who are critical of trace theories of memory. Squares has argued that memory is the retention of knowledge and that retention in general does not require a causal analysis. In the case of the retention of the colour of some indigo curtains, whilst it does follow that the curtains should have been indigo all along, it does not follow that the curtains should be indigo now because they were indigo before. It does not follow that there should be any causal connection between the previous and present colour-state of the curtains. In Squares' opinion, if memory traces are conceived of as a kind of inner causal chain, then like the colour in curtains, they can be retained without a trace[17] and '[e]ven if a causal chain bridges the gap it does not guarantee retention'.[18]

Norman Malcolm also argues that the concept of memory does not justify the requirement of a causal process. He also argues that memory does involve mnemic causation, a causation across a temporal gap. This phenomenon is an everyday occurrence in Malcolm's opinion - if I stoop to pick up a stone from my lawn and a nearby dog runs away, undoubtedly because someone had previously stoned the dog, then I presumably have a 'perfect' example of *mnemic causation*.[19] I do not agree: Malcolm's example of mnemic causation seems to be no more than a case of animal memory. In any case neither Square nor Malcolm have showed that a trace theory of memory, the most plausible causal theory of memory, is problematic - all that they have shown is that the concept of memory does not logically require a causal analysis for the intelligibility of 'memory-talk'.

Bursen however gives a general argument against the very coherence of trace theories in his book, *Dismantling the Memory Machine*,[20] after also arguing that the postulation of a memory trace is not conceptually inescapable but arises only from a tacit acceptance of mechanism. Bursen's argument against trace theories of memory can be neatly summarised by his criticism of the wax tablet metaphor employed by Plato in *Theaetetus*. For all trace theories, the trace must be isomorphic to what is remembered. Thus if I am remembering watching Plato walk down the street, then this must be structurally represented in the wax. Now possessing an image of Plato walking down the street will not constitute a memory unless I know that it is an image of Plato. To know this, I must recognise it as an image of Plato. But to do that I must already remember what Plato looks like. Now if this is done by an appeal to further images, then one falls into an infinite regress. On the other hand, if I can recognise the image without further images, then why can't I recognise Plato without use of any image at all?[21] Stated more generally, according to trace theory, a person recognises something if and only if there exists a matching trace. But how is the existence of a match itself recognised? If there is no homunculus in our head, then this can only be by means of a mechanism - but the recognition of the existence of a match presupposes memory on the part of the retrieval mechanism.

This is Bursen's most powerful argument against the trace theory of memory, but it is not ultimately successful. His critics have argued that the existence of a matching trace does not itself need to be recognised because it is a mistake to ascribe 'all of the cognitive features of the whole thinker to its parts'.[22] Another way of explaining this is by the following analogy. Recognition is like the opening of a door with a combination lock. Traces are like numbers fed into the lock. If the right number occurs, the door opens; if the right trace exists, recognition occurs. The 'right' trace, like the right combination of numbers, is simply the one that leads to recognition, or in the case of the door, an opened door.[23]

Braude is also concerned with refuting the thesis which he calls 'the myth of the internal mechanism' (MIM) and gives arguments which are more satisfactory than Bursen's. MIM is the thesis that it is possible to explain why a subject is in an occurrent or dispositional (intentional) mental state by reference to some physiological mechanism or structure which is either identical with or causally responsible for that mental state. Braude takes this thesis to be the principal metaphysical thesis upon which vast areas of modern cognitive psychology and the brain sciences rest, for they presuppose that mechanisms for at least some human cognitive abilities exist and are in principle discoverable. If the 'principle' of the internal mechanism' is false then 'numerous academic disciples would thus turn out to have no foundation'.[24]

Braude discusses the specific case of human memory, but believes that his arguments can be applied *mutatis mutandis* to other cognitive functions as well. He employs two arguments in support of this conclusion. The first argument is that the principle of the internal mechanism presupposes that there is not a one-many relationship between the types of psychological states and the types of mental states that may in turn be correlated with it. The second argument is that in order to identify some mental state m to be of type o, we suppose that there must be some specifiable set of properties in virtue of which a given state is of that type. However, providing necessary and sufficient conditions for a mental state to be of a specific type is in Braude's opinion a mistaken task. In the case of memory traces for example, a memory of a particular kind, say that of an old friend, may take an indefinite number of different forms having nothing in common apart from being a memory of an old friend. In remembering an old friend one may remember any number of things: a verbal description, image or event in which he/she featured. This remembering of some x need have nothing in common apart from being instances of x. The memory of x however is taken to explain an instance of remembering x. But if instances of remembering x do not necessarily have any important properties in common, the memory trace, serves no non-trivial explanatory purpose. If there is no common set of properties to link the trace to, the trace can serve no causal role. Virtually anything done by a subject in the right context may count as a remembering of x.[25] Even saying that there are different traces for different kinds of rememberings does not enable the trace theorist to escape this problem because these subsets of rememberings are not linked by any common set of properties '[i]f the members of some subset of rememberings . . . need have no relevant properties in common (save that of belonging to that subset), why suppose they are causally linked to the same thing?'[26] Thus for Braude changes in a person's cognitive abilities do not need to correspond to any internal structural changes in the brain at all: they are simply brute facts about the person. But facts about remembering are by no means miraculous; we may appeal to various empirically established regularities about the sort of things which I tend to remember, but that general ability is a fact about me for which no explanation in terms of memory traces exists or can exist.[27]

The conclusion to be drawn from the above criticisms is that the materialist account of memory is in trouble. Is there some way of saving trace theories from the perspective of a modified materialism? That is the question which I now wish to consider.

Memory and general resonance theory

Ninian Marshall[28] has presented a physical theory, devised to primarily explain extra-sensory perception, but which also can be applied to some problems in psychology, such as memory. Telepathy according to Marshall is the partial reproduction in one brain of a pattern in another brain. To account for this Marshall formulates a hypothesis which formally resembles Newton's Law of Gravitation which he calls the Hypothesis of Resonance:

> The Hypothesis of Resonance: 'Any two structures exert an influence on each other which tends to make them become more alike. The strength of this influence increases with the product of their complexity, and decreases with the difference between their patterns'.[29]

The influence spoken of in this hypothesis is called eidopoic influence and the change in pattern produced by an eidopoic influence 'increases with its strength, and with the quantum mechanically defined probability of the change'[30] This hypothesis, which formally resembles Newton's Law of Force, is called the Hypothesis of Eidopoic Influence. It claims that structures which are subject to eidopoic influence must have 'indeterminate' features in their observable behaviour. It is well known that this indeterminacy occurs in simple micro-physical systems, and it also occurs according to Marshall, in the response of nerve cells to a near-threshold stimulus. Marshall's theory then, has a number of novel features. It postulates action at a distance in space and time as well as coordinated action at many points in a structure.[31] Further, Marshall's theory has 'altogether discarded' the following properties of physical causes:

(1) the necessary involvement of a transmission of energy or momentum
(2) the cause acts before the effect
(3) a continuous chain of intervening events from cause to effect.

Telepathy is explained on Marshall's theory as resonance between two brains. According to the Hypothesis of Resonance, telepathy will be stronger between individuals with more closely structured brains, such as identical twins, than between individuals whose brain structure more widely differ.[32] Long term complex memory however is the result of a brain's resonance with a past state of itself; the more complex the memory pattern, the greater the influence of resonance. Long term memory then does not require a storage mechanism. Further Lashley's Law of Mass Action can be explained on this position because resonance influence does not presuppose that the same neurones are always involved in the recall of a specific memory. Any pattern in the cortex which resembled a previous pattern would be inclined to resemble it more over time. Telepathy and memory therefore are not radically different phenomena, but are both simply explained by the theory of resonance. Memory however is more reliable than telepathy. Two factors are responsible for this. First the resonance with one's own past brain states is of greater strength than resonance between different brains, because a brain pattern is more likely to resemble a previous pattern of its own, than a pattern of another brain.

Let us turn now to the task of criticising Marshall's theory. First this theory involves altogether discarding rather than modifying three fundamental properties of physical causation, as well as postulating action at a distance in space and time. Many materialists will find this problematic. Nevertheless, Marshall's theory of memory is not inconsistent with modified materialism. The materialist may include resonance and eidopoic influence in the class of material things. However, to do so is to admit that memory is not explicable to present physical theory, and most materialists believe that it is.

Second, it is doubtful whether the hypothesis of resonance is true for any structures. Marshall seems to have formulated his hypothesis with an explicit concern to explain telepathy and memory without much concern for various implications of his hypothesis. How would, for example two complex information processing systems such as a human brain and a computer become more alike, apart from the irrelevant sense of a possible increase in entropy over time? How would a red pen and a blue pen become more alike? How would a photon and an electron become more alike? Notice that these are not irrelevant questions because the Hypothesis of Resonance refers to *any* two structures one would wish to consider.

Third, Marshall's theory offers no mechanism by which structures can become more alike. It is hardly reasonable to say that this is simply a brute fact about reality because Marshall's theory involves a very significant departure from what we have taken to be the common properties of physical causes. Are eipodoic influences really physical influences rather than being paranormal? In his paper Marshall presents a mathematical analysis of the theory of resonance, but this does not show that his theory is really a physical theory. This fact is compatible with the views that paranormal phenomena can be analysed with the aid of mathematics. If we altogether discard the thesis that physical causes involve a transmission of energy or momentum, a view physicalists have taken to be basic to physical causation, then what right have we to call our theory a physical theory? The question remains: why should complex structures become more alike?

A general theory of life, influenced by the work of Marshall, has been proposed by Rupert Sheldrake.[33] His *A New Science of Life* contains both a detailed criticism of contemporary mechanistic biology as well as an attempt to go beyond this view point by advancing a new hypothesis based upon the notion of formative causation. Sheldrake is particularly concerned with the problem of *form*:

> Time after time when atoms come into existence electrons fill the same orbits around the nuclei; atoms repeatedly combine to give the same molecular forms; again and again molecules crystallise into the same spatial patterns; seeds of a given species give rise year after year to plants of the same appearance; generation after generation, spiders spin the same types of web. Forms come into being repeatedly, and each time each form is more or less the same. On this fact depends our ability to recognise, identify and name things.[34]

It may be thought that the problem of form can be disposed of rather quickly: fundamental physical principles and laws are taken to be explanatory prior to the actual forms of things. This view is mistaken on the two principal accounts of laws of nature available. If laws of nature involve, as Russell believed, constant conjunction of events, then form is not explained as we are seeking an explanation of *why* entities described in event descriptions stand in such a relationship in the first place. Appealing to more basic constant conjunctions of more basic events doesn't help because the form of these entities described in these more basic event descriptions must also be described. On a realist account of causality, form cannot be *ultimately* explicable by the action of basic generative mechanisms as these mechanisms will have a form which requires explanation. One could say of course that these natures do not require explanation, that their action is a brute fact, but this seems arbitrary as what is a brute fact today is often explained tomorrow.

The hypothesis of formative causation proposes that morphogenetic fields play a causal role in the development of all systems of varying forms of complexity. These fields are much like gravitational fields in that they causally interact with matter whilst they cannot be directly observed. Each kind of physical, chemical and biological system has its own characteristic form, and each its own characteristic morphogenetic field. Thus there is one for protons, one for electrons, another for various sorts of protein molecules and so on. Organisms are hierarchically organised at all levels of complexity, these systems being called *morphic units*. Higher level morphic units co-ordinate the arrangement of their component parts through the action of their morphogenetic fields on the morphogenetic fields of morphic units of a lower level. Various organised systems serve as morphogenetic germs around which a higher level morphic unit comes into existence with its specific morphogenetic field.

> The morphogenetic germ is a part of the system-to-be. Therefore part of the system's morphogenetic field corresponds to it. However, the rest of the field is not yet 'occupied' or 'filled out'; it contains the *virtual form* of the final system, which is actualised only when all its material parts have taken up their appropriate places. The morphogenetic field is then in coincidence with the actual form of the system.[35]

Morphogenesis involves the higher level field modifying the probability structure of events in the lower level morphic units, for example by restricting the number of possible arrangements which would be permitted by the probability structures of their constituent units. This form of explanation seeks to explain the behaviour or complex systems, not from the bottom up (i.e. reductionistically), but *holistically*. Further the action of morphogenetic fields are not fully explicable by quantum mechanics: 'There is no reason why the morphogenetic fields of atoms should be considered to have a privileged position in the order of nature; they are simply the fields of morphic units at one particular level of complexity'.[36] Nevertheless, not *all* form in living organisms is determined by formative causation. Some patterns may come about through random processes and the actions of physical forces as D'Arcy Thompson well described in his famous *On Growth and Form*.[37] Sheldrake maintains though that purely physical explanations of biological morphogenesis have had very little success in systematically accounting for the nature of biological form.

What determines the particular form of the morphogenetic field? The 'Platonic' conception of morphogenetic fields holds them to be eternal objects which are not explicable in terms of anything else and which can be neither created nor destroyed. Sheldrake's own view is that forms are repeated because of a causal influence from previous similar forms. The form taken by a system is not determined on the first occasion of its appearance. However once a specific form is taken by a system, similar systems will take on this same form because of trans-spatial and trans-temporal influences, that is, action at a distance across space and time otherwise the morphic influence of past systems is not attenuated by spatial and temporal separation, but is present everywhere. The possibility also exists that there may be morphic influences from future systems exerting causal influence backwards in time.[38]

It is a consequence of the hypothesis of formative causation, that the form of a system depends upon the cumulative morphic influence of previous similar systems, this influence becoming stronger with increasing repetitions. Now this hypothesis which was tailored to explain the constancy of form does so at the expense of failing to explain macro-evolution: how do, and why should stable forms evolve? Why should one general form of animal have evolved from another despite smaller adaptive changes, if one form has existed in a stable fashion for millions of years? Why is there the 'prodigious diversity of living organisms'.[39] If organisms are self-maintaining systems wouldn't one expect slight deviations in form to be overcome by the dominant morphic influences? Sheldrake concedes this in his discussion of the origin of new forms when he says that 'neither the repetition, modification, addition, subtraction nor permutation of existing morphogenetic fields can explain the origin of these fields themselves'.[40] Hence 'from the point of view of natural science, the question of evolutionary creativity can only be left open'.[41]

How is memory explicable by the hypothesis of formative causation? Sheldrake's view is that memories are not stored physically within the brain, but involve a 'direct action across time'.[42] He considers a number of hypotheses which would account for human memory, each of which depends upon a distinctive metaphysical position. Modified materialism regards morphogenetic fields and motor fields as aspects of matter. Conscious memory depends on morphic resonance from past states of the brain. Neither conscious nor unconscious memories need to be stored by traces within the brain. An alternative non-materialist view also discussed by Sheldrake is a dualistic interactionist account of the self. Here the conscious self interacts with motor fields associated with the body - these motor fields mediate the interaction between the mind and the body. The motor fields effect the brain much as morphogenetic fields effect the body but the conscious self is not identical to such motor fields. On this view conscious memories need not be stored materially in the brain or depend upon morphic resonance: the self may simply have direct access to its own past states simply because of similarity with present states.

I have commented in a previous footnote about some of the problems with the idea of direct action across time, and I do not believe that Sheldrake has solved the problem of memory. Nevertheless our discussion has illustrated the major problems facing materialist accounts of memory. Even if the materialist accepts Sheldrake's position of modified materialism there is another problem which faces him/her which we shall now turn to.

The correspondence thesis

The correspondence thesis or principle of psycho-physical isomorphism holds that psychological phenomena thoughts, emotions and images, correspond in a one-to-one fashion with certain brain states and processes. This thesis is taken by both materialists as well as supporters of alternative theories such as dualistic interactionism, parallelism and epiphenomenalism to be at the very least a highly plausible factual hypothesis. If the correspondence thesis was undermined, then in the words of one of its critics, so would be undermined 'the presuppositions of much of the current (mind-brain) debate'.[43] In this section I wish to consider the arguments of some contemporary critics of the correspondence thesis and discuss how these arguments serve to undermine the present mind-brain problem.

Critics of the correspondence thesis include Goldberg[44], Solomon,[45] Lurie,[46] Braude[47] and Puccetti and Dykes.[48] Here there is space only to discuss Solomon Puccetti and Dykes, Lurie and Braude and our discussion of this complex material must itself be highly selective.

Robert Solomon was one of the first philosophers to question the correspondence thesis on empirical grounds. He presented a useful discussion of neurological research that concluded with the bleak news for materialists that neurology, including the brain stimulation work of Wilder Penfield 'has waged a war against just those models upon which philosophers have placed their faith in "future neurological researches"'.[49]

A similar conclusion was reached by Puccetti and Dykes. They asked whether it is possible to explain the subjective differences between seeing, hearing and feeling something by inspecting the structure and function of the visual, auditory and somesthetic cortex. If this was so, then it would be possible to find theoretically significant neural correlates of subjective experience so that different qualities of experience are explicable by reference to the physical structure of the brain. Their conclusion was that 'one cannot explain the subjective differences between sensory modalities in terms of present day neuroscientific knowledge. Nor do present trends in research provide grounds for optimism'.[50] They voiced the following objections to the classic mind-brain reductivist program. First the principles of electrochemical excitation apply right across cortical tissue and cannot 'distinguish one sensory projection area from another'.[51] Second, looking 'deeply' at the activities of cells of the cortex gives us no insights, for even in the case of electrical activity, there is no evidence of a substantial difference between cells in the somesthetic, auditory or visual cortex. These and other more technical neurophysiological considerations lead them to conclude that we are up against an 'impenetrable mind-brain barrier'.[52] It will be beyond the scope of this essay to consider the objections made to Puccetti and Dykes' position beyond nothing at the majority of the reviewers are critical, but they all differ on what is wrong with their argument! This suggests that Puccetti and Dykes' position is a fruitful one, that may be open to further defence. If this proves to be so, then the philosophical attacks upon the correspondence thesis will be strengthened even further and the mind-brain reductivist material program be made even more implausible.

Braude also rejects the correspondence thesis. He distinguishes between two general forms of the material mind-body identity theory. The first form holds that there are, could, or will be discovered lawlike correlations between types of mental states and types of physical states. This position has been abandoned by many contemporary materialists in favour of a token identity theory. According to this position, every mental state-token is a brain state-token, but there need not be any lawlike correlation between brain state-types and mental state-types. This position was defended by Davidson in a number of papers under the name of *anomalous monism*.[53] In recent times other philosophers have argued against mind-brain reductions by bridge laws but have accepted the idea of a token-token reduction of mental events (or states) to brain events (or states).[54]

Braude has given an argument that anomalous monism, rather than escaping the difficulties of type-type versions of the identity theory, actually presupposes type-type correlations. This means that if the type-type identify theory is wrong, then so must the token-token identity theory be wrong. Braude argues that mental states can only be characterised *positionally*, with respect to the way in which certain events fit into a sequence of events. For example, the same mental image in different contexts may represent different things. Anomalous monism maintains that a subject's mental state is a particular brain state. But if a given brain state corresponding to a certain mental image (say of a dog) can represent different things in different contexts (such as some thought about either a specific dog or dogs in general), then the brain state represents no more than the image itself and a particular mental state in a subject cannot be *a* particular physiological state.

However, couldn't anomalous monism be modified to maintain that the mapping between mental state or mental event, and a brain state or brain event is not one to one, but many-one, that is that the brain state o corresponding to the subject's mental image may represent different things in different contexts, but still, the mental state and its context are both material? Braude doesn't discuss this objection but perhaps he does not need to in the light of a more general objection to anomalous monism. If predicate expressions of the 'language used to describe mental events' need not map onto predicate expressions of the 'language used to describe brain states', o, then there is no reason to believe that those referring expressions map onto those of o, that is the objects to which the predicate expressions apply. As he puts it:

> If physiology and psychology distinguish different kinds of phenomena - if they do not divide the world into kinds or types, in anything like the same way - they may also have entirely different criteria of individuation within their respective domains; in that case, their domains may not divide up into members, or tokens, in anything like the same way.[55]

It is concluded that the anomalous monism assertion of token-token identity is essentially arbitrary. It is of course still possible to maintain that if a subject is in a mental state token m, then there is some brain state token b such that m could not exist without b. This claim however is no longer a statement of identity, rather it is a statement of causal dependence and is consistent with a dualistic position such as epiphenomenalism. A retreat to such a thesis therefore can offer no comfort to the materialist. Second if there are radically different criteria of individuation for the mental and the physical, even if m could not exist without b, how could we possibly know it?

Yuval Lurie has also given a radical critique of the correspondence hypothesis, arguing that it is logically incoherent. This incoherence arises from the psychological side and therefore cannot be removed by a shift from a one-one correspondence to a many-to-one (brain states - psychological phenomena). His argument involves selecting a particular belief such as the belief that the central library to Ben-Gurion University will be closed next Monday, and showing that this belief is not a discrete phenomenon with clearly drawn boundaries. Hence 'the brain state in question not only turns out to correspond to many beliefs in each case considered, in each of these cases it corresponds to a different set of an undeterminable number of beliefs which all amount to having a belief that the library will be closed next Monday'.[56] He argues in support of his claim that this belief is not discrete, by first pointing out that the number of beliefs which people may acquire alongside their belief that the central library will be closed on Monday is not determinate because there are an indefinite number of other beliefs that may also be acquired. As the number of beliefs is not determinate Lurie shows that it is not possible to eliminate these extra beliefs by some inductive procedure. This leads him to conclude that psychological phenomena do not possess a determinate essence as is required for the cogency of the correspondence hypothesis.

Conclusion

I have surveyed some radical challenges to the physicalist reductionist account of the nature of the mind. It obviously has not been possible to discuss in any detail anticipated physicalist responses to these arguments consequently any conclusion drawn about the cogency of physicalism must be tentative. Nevertheless, it is clear that there is an emerging critical research program of physiologically informed philosophers who are united in their belief that human persons are not mere mechanisms and that the human mind is not reducible to the human brain. I conjecture that future battles between physicalists and anti-physicalists may take place on the terrain which we have surveyed.

Notes

1   For a discussion of the metaphysical and metascientific ramifications of the physicalist reductionist program see my *Reductionism and Cultural Being*, (Martinus Nijhoff, The Hague, 1984)

2   c.f. H. Bergson, *Matter and Memory* (George Allen and Unwin, London, 1911; 7th Impression 1962) and *Mind-Energy: Lectures and Essays* (Macmillan, London, 1920).

3   We are concerned here only with human memory, rather than the more general problem of *inanimate memory*. Nevertheless it is worth noting that in the recent literature the problem of inanimate memory has been raised. Pre-Victorian physics ascribed memory to lumps of iron. Today we talk of *hysteresis*, a piece of unmagnetized iron placed in a magnetic field that is of increasing strength will increase in magnetic strength less quickly than the field, but then the rate of increase becomes greater. Magnetization grows on an S-shaped curve and also decreases on an S-shaped curve. Speaking anthropomorphically the iron seems to 'remember its past history'. I.D. Mayergoyz 'Mathematical Models of Hysteresis', *Physical Review Letters* vol.56, no.15, 1986, pp.1518-1521 has given a mathematical description of hysteresis, giving the necessary and sufficient conditions for the representation of actual hysteresis non-linearities. J. Maddox 'Is There Inanimate Memory', *Nature*, vol.321, 1986, p.11 speculates that Mayergoyz's equations may 'explain some of the properties of memory as people know it'. I doubt however whether Mayergoyz's model explains hysteresis. He says that his model whilst mathematically general is 'phenomenological' in nature (p.1518). It is one thing to mathematically represent a physical process, quite another to show why a hysteresis curve can be viewed as a function of iron's constitution. The latter question requires a *physical* theory.

4   c.f. E.R. John, *Mechanisms of Memory* (Academic Press, New York, 1967), p.203.

5   C.B. Martin and M. Deutscher, 'Remembering', *Philosophical Review*, vol.75, 1966, pp.161-196. Citation p.189.

6   For a critical discussion of the problem of memory from a physiological perspective c.f. H.A. Buchtel and G. Gerlucchi, 'Learning and Memory in the Nervous System', in R. Duncan and M. Weston-Smith (eds) *The Encyclopaedia of Ignorance* (Pergamon Press, Oxford, 1977), pp.283-297.

7   N. Marshall, 'ESP and Memory: A Physical Theory', *British Journal for the Philosophy of Science*, vol.10, 1960, pp.265-286.

8   ibid pp.281-282.

9   J. Beloff, 'Is Normal Memory a Paranormal Phenomenon?', *Theoria to Theory*, vol.14, 1980, pp.145-162. Beloff himself attempts to solve the problem of memory by combining a trace theory of storage with a paranormal theory of retrieval. This is at best only a 'partial' materialist account of memory and in any case Beloff is sympathetic to the possibility of the existence of extra-cerebral memories where no brain traces could store memories, because in a life-after-death state, the original brain no longer exists. If Beloff is right, matters have not improved substantially for the materialist.

10 W. Penfield, *The Excitable Cortex in Conscious Man*, (Liverpool University Press, Liverpool, 1958).

11 Even in animals authorities have concluded that behaviour evoked by brain stimulation cannot be accurately predicted by knowledge of the location of electrodes. c.f. E.S. Valenstein, *Brain Control: A Critical Examination of Brain Stimulation and Psychosurgery* (John Wiley, New York, 1973).

12 Bergson, op.cit., note 2, p.231.

13 S.E. Braude, *ESP and Psychokinesis: A Philosophical Examination*, (Temple University Press, Philadelphia, 1979), p.201.

14 It is worth mentioning here that there has been at least one attempt to justify trace theory *a priori*. Deborah Rosen in her 'An Argument for the Logical Notion of a Memory Trace', *Philosophy of Science*, vol.42, 1975, pp.1-10, argues that one can show deductively from reasonable empirical premises the conclusion that between a remembering event and a prior learning event there is at least one spatio-temporal causal chain, which is by definition a memory trace. This idea of a memory trace she calls, the logical notion of the memory trace. Her argument though makes use of the principle of spatial and temporal contiguity 'that there can be no direct causes remote in space or time' (p.7). She points out that it is this principle which has led to the postulation of the memory trace (p.8). Unfortunately however, this argument begs the very question at issue, which is whether or not memory is a counter-example to the principle of spatial and temporal contiguity.

15 R. Squires, 'Memory Unchained', *Philosophical Review*, vol.78, 1969, pp.178-196.

16 N. Malcolm, *Memory and Mind*, (Cornell University Press, Ithaca and London, 1977). On mnemic causation c.f. E. Rignano, *Biological Memory*, (Kegan Paul, Trench, Trubner and Co., New York, 1926) and On the Inheritance of Acquired Characters (Open Court, London, 1911). Bertrand Russell in *The Analysis of Mind*, (George Allen and Unwin, London, 1921), pp.77-92 also recognised that mnemic causation may prove to be an ultimate explanation for memory. This was not a problem for Russell because for an empiricist causation is merely observed uniformities, of sequence. Realists find causation across a temporal gap problematic. Causation involves a generative mechanism and everything constituting a memory-belief is happening now. But the past no longer exists, so mnemic causation is impossible because the cause does not exist. How can that which doesn't exist be a cause of something which does exist?

17 Squires, op.cit., note 15, p.196.

18 ibid, p.180.

19 Malcolm, op.cit., note 16, pp.187-188.

20 H.A. Bursen, *Dismantling the Memory Machine: A Philosophical Investigation of Machine Theories of Memory*, (D. Reidel, Dordrecht, 1978).

21 ibid, pp.109-110.

22 'Review of Bursen, *Dismantling the Memory Machine*' *Review Metaphysics*, vol. XXXV, 1982, pp.859-860. Citation p.860.

23 c.f. also C. Mortensen, 'Review of Bursen, *Dismantling the Memory Machine*' *Australasian Journal of Philosophy*, vol. 59, 1981, pp.130-133.

24 Braude, op.cit., note 13, p.187.

25 ibid, p.192.

26 ibid, p.192.

27 Braude also presents an insightful criticism of the recent theoretical interest in developing holographic models of memory; he argues that these theories are nothing more than trace theories and as such stand open to the theoretical problems of trace theory.

28 Marshall, op.cit., note 7.

29 ibid, p.266.

30 ibid, p.267.

31 ibid, p.268.

32 ibid, p.280.

33 R. Sheldrake, *A New Science of Life: The Hypothesis of Formative Causation*, (Blond and Briggs, London, 1981).

34 ibid, p.92.

35 ibid, p.76.

36 ibid, p.85.

37 D.W. Thompson, *On Growth and Form*, (Cambridge University Press, Cambridge, 1942).

38 Sheldrake, op.cit., note 33, p.96.

39 ibid, p.139.

40 ibid, p.149.

41 ibid, p.150.

42 ibid, p.47.

43 R.C. Solomon, 'Doubts about the Correlation Thesis', *British Journal for the Philosophy of Science*, vol.26, 1975, pp.27-39. Citation p.27.

44 B. Goldberg, 'The Correspondence Hypothesis', *Philosophical Review*, vol.77, 1968, pp.438-454.

45 Solomon, op.cit., note 43.

46 Y. Lurie, *The Correspondence Thesis* (PhD Thesis, Cornell University, 1973) and 'Correlating Brain States with Psychological Phenomena', *Australasian Journal of Philosophy*, vol.57, no.2, June 1979, pp.135-144.

47 Braude, op.cit., note 13.

48 R. Puccetti, and R.W. Dykes, 'Sensory Cortex and the Mind-Brain Problem', *Behavioural and Brain Sciences*, vol.3, 1978, pp.337-375.

49 Solomon, op.cit., note 43, p.33.

50 Puccetti and Dykes, op.cit., note 48, p.337.

51 ibid, p.340.

52 ibid, p.341.

53 c.f. D. Davidson, *Essays on Actions and Events*, (Oxford University Press, Oxford, 1980).

54 e.g. J.A. Fodor, 'Special Science (or: The Disunity of Science as a Working Hypothesis)', *Synthese*, vol.28, 1974, pp.97-115.

55 Braude, op.cit., note 13, p.181.

56 Lurie, op.cit., note 46 (article) p.143.

5 Physics and existentialist phenomenology

Robert C. Trundle

Introduction

I will argue that existential phenomenology is not the puerile movement pictured by neo-positivistic philosophers of science. Such philosophers, having arrogated to themselves the definition of what philosophy is, have isolated themselves from fruitful insights. Thus they are hopelessly entangled in knotty epistemological problems that become most poignantly evident in their construals of physics.

Since the corroboration of theories in physics hinges on observation, it is not surprising that central controversies involve their received-view appeals to infallible 'protocol sentences' based on observation. The challenge to such observation, in terms of relativistic theses that it is theory-dependent, has generated such responses as a *reductio ad absurdum* rejection of theory-dependency theses (F. Suppe and W.H. Newton-Smith) and an observation-revisability thesis wherein instrument-aided observation together with theoretical experimentation increasingly ameliorate naked-eye observation (Alan Chalmers). But besides arguing that naked-eye observation is necessary for manipulating the instruments that paradoxically ameliorate it and that *reductio ad absurdum* strategies fail to establish observational truth, I shall argue that such truth has its ontological basis in an observational consciousness of which the physicist is also phenomenologically conscious.

Specifically, I will seek to show that the phenomenological consciousness of observation belies the fact that such observation is laden with features which phenomena have *in themselves* - independently of observation-theoretical concepts. While this does not obviate a theory-laden thesis of observation that permits different theoretical interpretations, it provides an ontological anchor for articulating their success and truth. And in addition to the provision for truth, the physicist's consciousness of his observations and predictions of phenomena comprises his very freedom to engage or not engage in scientific inquiry. Importantly, this permits responsibility and a moral dimension for theoretical physics that Kant's influential *Critique* strictly relegated to 'practical' matters in virtue of freedom being a 'noumenon' as opposed to a 'phenomenon.' But the phenomenology of consciousness indicates that a determinism endemic to predictability is phenomenologically and conceptually connected to freedom.

My phenomenological analysis of physics *inter alia* in terms of freedom and determinism is, hopefully, a significant contribution. For it is the 'softer' human sciences that have been typically addressed by existential phenomenology. Thus the following discussion is not intended as an exegesis of the thought of Sartre, Merleau-Ponty, Heidegger and so forth. Rather, while it elicits their thought, it has the unusual aim of strengthening an observational basis for truth-claims of physics; one that, in virtue of relating observation to consciousness, connects determinism to freedom and the latter to truth.

Truth and observational consciousness

Phenomenology emphasizes that the observational consciousness of physicists is complemented by their being incontrovertibly aware of what they observe. Further, what they observe may be understood as both a phenomenon laden with concepts and a concept-independent 'phenomenon-in-itself.' This obtains by virtue of observational consciousness not being exhaustively entangled with observation-theoretical concepts. And this is an important point because it is precisely a partial independence from such concepts that yields limited ascriptions of reality to observational referents and truth or limited truth to the theoretical descriptions in terms of which they are understood.

If we consider various experimental contexts - or comments on experiments by physicists, the disregarded significance of observational consciousness and awareness of it becomes clear. Thus, for example, consider experimental setups in the classic case of Heisenberg's uncertainty principle. A professor of physics, Harvey E. White, notes that, according to Niels Bohr, this principle provides complementary descriptions of the same phenomenon in terms of wave and corpuscular properties of matter. Since, however, it is impossible to design an experimental setup that simultaneously shows both properties, a given experiment reveals details of either the wave or corpuscular character 'according to the purpose for which the experiment was designed.'[1]

On the one hand phenomenology draws attention to the fact that there could be no intelligible talk of a *purpose* in designing an experiment independently of the physicist being incontrovertibly aware of what he observes as well as of his intention to observe it in a certain manner. On the other hand his intention to observe the phenomenon in terms of wave or corpuscular characteristics is senseless apart from the phenomenon having *in itself* distinctive characteristics. If the observation of such characteristics is understood in terms of the theory-dependency thesis, then the observation is no less conceptually bifurcated from phenomena than the concept- dependent observation of the scintillation counters, slits, and screens of experimental apparatus that are employed to produce and measure the phenomena.

This is poignantly relevant to Alan Chalmers' unqualified assertion, in *Science and Its Fabrication* (1990), that 'What is correct about the "theory-dependence of observation" thesis is not that observation in science lacks objectivity, but that the adequacy and relevance of observation reports within science is subject to revision.'[2] For if such revision unqualifiedly holds, say in terms of Galileo's telescopic observation removing the irradiation of naked-eye observation, then the naked-eye observation of experimental instruments would itself be subject to continual revision. In other words, phenomenology underscores that a weak or strong thesis of theory- dependency has tended to conflate observation *per se* with implicit theoretical judgments about it.

In exactly this sense phenomenology distinguishes the 'naked-eye observation' endemic to irradiation (wherein, say, the size of Venus varies), from the instrument-aided observation of its invariance. Phenomenology indicates that there is nothing wrong with naked-eye observation per se, whether of Venus or fringe patterns in an experimental setup in quantum mechanics. What is wrong is a simple untested theoretical judgment that might be implicitly incorporated into the unaided observation. An untested judgment about naked-eye observation should be superseded by a tested one whose function is precisely to explain an unaided observation in question.

It is interesting in this respect that St. Augustine, who is generally credited with having anticipated many insights of existentialism,[3] lends his thought to the notion that there is nothing wrong with a *naked-eye observation* of an oar appearing bent in the water. That is how it should appear. What is wrong is the *judgment* that it is really bent. It is exactly the function of theory, say one of optics, to explain if not predict the observation.

If naked-eye observation was exhaustively skewed by theoretical concepts, we could not make sense of the possibility that the physicist may be observationally conscious of such things as a 'wave phenomenon' or 'celestial body' without thinking about them at all. This might occur in such mundane contexts as when the physicist is having lunch or is bored; hence the epistemological significance of the existential emphasis on mood. There is no more reason why he will observe a wave phenomenon, glowing patch of a cathode-ray tube, or light sources in a dark celestial background (irradiation) as scientific phenomena on these 'mood-laden' occasions than he will observe the patterns on a television screen as persons or particular things when he is bored at home.

The epistemological and ontological significance of what he observes at home are as reliant on his freely chosen intentions as what he observes in the laboratory. At the same time unless what he observes in the laboratory or elsewhere had distinctive features in themselves apart from his intentions, it might be his wish, will, or thought that determined what he observed. In this sense existential phenomenology lends itself to the notion that observation involves a direct cognitive or noncognitive consciousness of something other than consciousness together with an indirect phenomenological awareness of such consciousness.

Thus 'consciousness' is another way of understanding the observations of which the physicist is implicitly conscious. Unless there were such an implicit consciousness or awareness, there would be only a direct observational consciousness of something without any awareness of it. But the physicist in this case would be unable to distinguish his observation of something from the thing of which he is observationally conscious. And hence he could not intelligibly speak of his being conscious of anything at all.

It is noteworthy that there is no more reason why persons must think at all times when they are conscious than they must speak when they are conscious. But whenever they speak or think they are implicitly conscious of their speaking or thinking. Thus while there can be no cognitive or conceptual activity - including concept-laden observational consciousness - without an indirect consciousness of it, there can be consciousness per se without such conceptual activity. Hence the conceptual activity that accompanies observational consciousness has its ontological roots in a nonconceptual observational consciousness of something of which one is implicitly conscious. And therefore while physicists may observe some phenomenon as a wave or celestial body in the context of theoretical experimentation, their observations are *laden* with a nonconceptual consciousness of the *phenomenon-in-itself* as well as with observation-theoretical concepts.

It is unreasonable to suppose that there is observational consciousness of phenomena-in-themselves independently of concepts but that concept-laden observation of phenomena is exhaustively bifurcated from it. But this raises the possible conflation of the phenomenological phenomenon-in-itself with the Kantian thing-in-itself.

Observational consciousness and Kant's thing-in-itself

I shall argue that observational consciousness of phenomena-in- themselves is the basis for scientific objectivity after briefly articulating the pejorative meaning with which these words or similar ones have come to be associated. For they tend to be associated with the 'thing-in-itself' (*ding an sich*) of Kant's *Critique of Pure Reason*.

Kant's Critique is connected with his ostensively critical distinction between reality as it is cognitively interpreted *(phenomena)* and as it is *in itself* apart from such interpretation *(noumena)*. Whereas noumena was in principle unknowable because it was not an interpretable 'raw material of experience,' phenomena was knowable in virtue of such experience being interpreted by an *a priori cognition*. The obvious incentive for Kant to appeal to such a cognition was that it seemingly transposed the Humean problem of causal determinism not being known a *priori* to an *a priori* cognition in terms of which a raw material of experience was interpreted causally.

This was important because the judgment that all events have causes was a necessary presupposition prima facie for the intelligibility of scientific inquiry. But besides the fact that such inquiry - as construed by Kant - ignored the incontrovertible reality of human freedom (because it was not a raw *material* of experience), Kant's thesis of an a *priori* cognition was no more empirically or logically knowable than the truth-valueless metaphysical principle of causal determinism. This meant that Kant exacerbated rather than ameliorated Hume's skeptical empiricism. For the truth-claims of physics could have no more metaphysical significance in describing what reality is really like than the truth-valueless metaphysical principle or judgment that such truth-claims presupposed.

But it is also generally disregarded that Kant's thesis of an *a priori cognition* also bifurcated cognitive observation from the very thing observed. For as Hume had ideas corresponding to mere sense impressions, Kant strictly had phenomenal ideas corresponding to a 'raw material of *sense experience.*' Notwithstanding such experience being interpreted by an a priori structure of mind, the mind's phenomenal ideas could be said to correspond to a *material* reality only on pain of Kant embracing a rationalistic dogma he disparaged.

Having said this, it does not take much imagination to see how Kant's notion of phenomenal ideas being *cognition-dependent* was transposed into the notion of their being *theory-dependent* by such philosophers of physics as Thomas Kuhn and Paul Feyerabend. The positions of Feyerabend and Kuhn are sometimes called *Weltanschauungen Analyses* because the theories that interpret phenomena a priori are themselves sometimes construed as being interpreted or determined a priori by different and possibly inconsistent worldviews of different cultures.

The rational post-Kantian community of Anglo-American philosophers have had more success in rejecting theory- *(Weltanschauungen-)* dependency theses than in articulating *how* there can be any concept-independent observational touchstone with reality. This is evident both in W.H. Newton- Smith's *The Rationality of Science* (1981) and Chalmers' previously noted *Science and Its Fabrication.*

Although Chalmers seeks to mediate between a positivistic verificationism and relativistic skepticism endemic to theory- dependency, Newton-Smith rejects the thesis of theory-dependency as logically incoherent and affirms a 'general faith' in observational truth. Neither philosopher, however, explicates how instrument-aided or unaided observation can have any touchstone with empirical reality without circularly employing the observation-theoretical concepts in question; the same question-begging difficulty to which Hume drew attention. Let me articulate such a touchstone after briefly examining the position of Newton-Smith and reiterating that of Chalmers.

Chalmers is correct in criticizing positivism insofar as it merely 'aimed to show that legitimate science is "verified," shown to be true or probably true by reference to "protocol sentences," facts revealed to careful observers by way of their senses.'[4] But Chalmers' unqualified appeal to revisable observational experience, on the basis of increasingly instrument-aided observation, begs the Humean question regarding how *instrument-aided observation* can be shown to correspond to empirical reality independently of the observation- theoretical concepts in question. Moreover, as previously mentioned, the strategy of revisable naked-eye observation raises the question of whether *naked-eye observation of instruments,* used in revising naked-eye observation of phenomena, is itself revisable by instrument-aided observation *ad infinitum.*

This is not an academic question whose significance is erased by everyday common sense. It is exactly the reliability of common- sense observation that begs for a phenomenological explanation that, not appealing to mere theory or metaphysics, appeals to the integrity of the common person's incontrovertible conscious experience. It is the same or similar experience that is inescapably connected to the physicist's pre-axiomatized or pre- formalized observational descriptions.

Newton-Smith made a more convincing case for such descriptions when he tacitly countered Chalmers' revisability strategy almost a decade before the latter's *Science and Its Fabrication.*

Thus on the one hand Newton-Smith acknowledges Chalmers' strategy of revision when he concedes that 'prior to the development of modern astronomy, anyone would have said that the stars were yellow,' but that 'on the basis of an instrument- aided study of stars we reject those observation reports.'[5] On the other hand he notes that 'we must not make the fallacious inference that, as any observation report no matter how well- corroborated may be rejected by appeal to theory, *all* observation reports might be rejected by appeal to theory.'[6] The challenge to all unaided observation reports on the basis of instrument-aided observation draws attention to the fact that Chalmers' strategy of observational revision does surreptitiously embrace the theory-dependency he rejects.

However, one hastens to add that Newton-Smith's rejection of theory-dependency, notwithstanding his 'general faith in the low- level O-reports we are inclined to make,'[7] does not *establish how* observation reports do veritably describe the reality addressed by physics or astrophysics. His notion that the latter cannot be coherently based on theory-dependent relativisms belies the *reductio ad absurdum* nature of his position. It is important to unpack his position in this sense because, while it is logically devastating to relativism, it might be mistakenly construed to be devastating to an observational ambiguity fostered by phenomenology as well.

I will discuss why a phenomenological ambiguity is not obviated by Newton-Smith's criticism after examining this criticism as well as features endemic to phenomena-in-themselves and a logico- mathematical schema. For it is such a schema that, appealing to an observational consciousness of phenomena-in- themselves, belies how inexact observational descriptions are reduced to exact theoretical ones. This gives rise to the notion that a given observational datum may be ambiguously explicated by two or more theories.

The relativistic picture of theory- (neo-Kantian Weltanschauungen-) dependency, that Newton-Smith challenges, holds that truth changes from age to age, theory to theory, culture to culture etc. This picture may be formulated in terms of the assertion, 'It is possible that sentence "S" is true in Θ and false in Ψ,' where 'Θ' and 'Ψ' refer to whatever it is to which truth is relative.[8] The assertion may mean that 'S' has different meanings in different ages, theories, or cultures. But if all the assertion means is that 'S' with different meanings has different truth-values in virtue of different truth- conditions, and if this is necessarily the case, then the assertion is trivially true.

But if 'S' in Θ and Ψ expresses the same sentence, say as understood in terms of proposition p, then the assertion is incoherent. For if 'S' in Θ and Ψ expresses the same proposition (p), then 'S' in Θ and Ψ has the same truth- conditions for what makes it true or false. And if the conditions for what makes 'S' true or false are identical in Θ and Ψ, then it is incoherent to assert that p *qua* S might be true in Θ but false in Ψ. It is, that is, a necessary condition for the sentence to express the same thing that it have the same truth-conditions for what makes it true.

I shall examine below why Newton-Smith's analysis of truth does not conflict with a phenomenological ambiguity wherein reality 'overflows' any univocal truth-claim about it. It is important to note at this point that the analysis does not establish a tenable observational base for theoretical truth- claims of physics. His notion of a realism of physics, wherein 'the sentences of scientific theories are true or false... in virtue of how the world is independently of ourselves,'[9] only underscores the need for a nonconceptual (culturally neutral) element in observation which renders intelligible the communication of different cultures or *Weltanschauungen*.

That *Weltanschauungen* Analyses are unhappily linked to relativistic theses of theory-dependency which had their origin in Kant is evidenced by Frederick Suppe's assertion, in *The Structure of Scientific Theories* (1979), that:

> Accordingly it [a Weltanschauung Analysis] conveniently can be viewed as a kind of neo- Kantian pragmatic position; unlike the nineteenth-century neo-Kantian philosophy of science..., this approach does not claim there is a unique set of categories determining the *Weltanschauung*, but allows that significantly different ones are possible...[10]

Whereas Kant's a priori cognition (with its categories of quantity, quality, relation, and modality) was indigenous to a universal human cognition, the *Weltanschauungen* analyses rendered cognitive interpretations relative to possibly inconsistent worldviews of different cultures.

Is it not possible that members of different cultures share in a noncognitive human consciousness that reveals rather than conceals what is subject to different but veritable observational interpretations? But an ontological element that yields objectivity or approximate objectivity to the observational data and testing of theories could not be explicable unless phenomena- in-themselves had general features that physicists experience.

From Kant's thing-in-itself to phenomenological experience

The experiential basis of physics can be fruitfully articulated in terms of John Compton's analysis of our phenomenological experience in 'Natural Science and the Experience of Nature' (*Phenomenology in America*[11]).

Such experience is elucidated by his reference to phenomena being characterized by their having *independence, individuality, extensiveness, and continuity*. But whereas, for Compton, such features exhaustively attach to *conceptual* pre-scientific experience, I argue that they are applicable to a *nonconceptual* observational consciousness as well. For unless Compton's phenomenological but conceptual experience is explicated in terms of noncognitive experience, phenomenology will be pejoratively associated with a paradigmatic (Kantian) conception of 'phenomena.' The latter, being cognition-dependent (via an a priori cognition), has peregrinated into construals of observation being theory-dependent.

Again, the thesis of theory-dependency is avoided by the notion of there being an observational consciousness of phenomena-in- themselves. And this notion does not invoke the Kantian thesis of possible noumenal realities behind phenomenal appearances. Rather it is connected with the physicist's immediate, incontrovertible, and phenomenological awareness of his observational consciousness apart from conceptual or cognitive activity. When, for instance, physicists observe phenomena, they are noncognitively conscious of 'something' *being there* independently of their will, wish, or thought. And the noncognitive or nonconceptual consciousness of something's being *independent* of thought is indisputably infused in the physicist's concept-laden observation. Hence scientific observations of phenomena are laden with an observational consciousness of phenomena-in-themselves.

The *independence* of phenomena-in-themselves is presupposed for the intelligibility of attempts to corroborate theories. For while well-established theories render reliable predictions about phenomena, the latter are sometimes 'surprising.' [12] If they were not thus surprising *inter alia* in virtue of what they are in themselves independently of the physicist's will, wish, or thought, the physicist's predictions would have no risk. It is precisely such risk on which both the integrity of corroboration depends and well-established theories may be challenged.

Thus, for example, Peter Beckman, Professor Emeritus at the University of Colorado at Boulder, has recently 'proposed a rival theory of physics which... fits the known facts and explains them more simply than Einstein's.'[13] Beckman's *Einstein Plus Two* (1987) embraced theoretical notions that were surprisingly supported by the experiments of Howard Haden, Professor of Physics at the University of Connecticut, in 1988. It would, according to Haden, allow us to bring back space and time as absolutes, and generally restore the classical world view of Isaac Newton.

The possible restoration of a Newtonian view is not being referred to here to suggest that an Einsteinian view is incorrect. Whether it is correct or not, the mere challenge to it by reputable scientists makes physics exciting and belies the surprises that attach to the *independence* of phenomena (by virtue of what they are *in themselves*). Such independence is further illustrated by two extraordinary events in 1989: the discovery of a pulsar spinning at a putatively impossible speed and 'cold fusion.'

The spinning pulsar, located at the heart of the spectacular star explosion known as Supernova 1987A, was calculated to be spinning 2,000 times per second. This rate of spin, being approximately three times faster than any other previously discovered, could only fail to fly apart by being dense enough to be a black hole. Whereas, however, a black hole is observationally invisible notwithstanding ostensively indirect instrument-aided detection, the surprising anomaly of the pulsar was more directly detectable. Thus Professor Stanley

Woosley, astrophysicist at the University of Santa Cruz, asserted that 'We have absolutely incontrovertible evidence of something that cannot exist [in terms of well-corroborated theories]... We like being surprised more than anything else - even more than being right.'[14]

The tension between being right in terms of established theories and surprises is further illustrated by the putative phenomenon of 'cold fusion.' The highly publicized experiment of generating heat by simply inserting a palladium rod in hydrogen-laden 'heavy water' ('cold fusion') was inescapably a surprise. For hundreds of millions of dollars in research has failed to produce 'hot fusion.' And the significance of this inexpensive experimental response to such expensive fusion obtains whether or not 'cold fusion' is experimentally corroborated in the sense of being replicated.

The very fact that physicists were surprised, without dismissing it out of hand, reveals their phenomenological experience of phenomena-in-themselves being independent of their will, wish, or thought. Indeed, in this case, it is likely that they would will or wish that there be experimental duplication. But this only further belies the fact that such a duplication is independent of the human will one way or the other. This is an important point in view of Sartre's 'coefficient of resistance' wherein our phenomenological awareness of our will increases in direct proportion to a resistance against it. For this underscores an existential notion of freedom in which freedom is not the mere ability to do what one wills but rather to will to do it even when one cannot apparently do it; a distinction of practical significance which attaches to everyday perseverance as well as scientific struggles to explicate and predict the 'impossible.'

It is beyond my purpose to expand upon the many existential and phenomenological notions connected to *independence*. It is important to note that such independence is conceptually and phenomenologically connected to that of *individuality*. Individuality, at the conceptual level, is articulable in terms of phenomena having a boundary or some coordinated principle of parts or action.[15] This permits identification. Thus it is noted by Compton that the scientific concern for invariance under stipulated transformations is a manifestation of identification.[16] But it is also important to note that such identification has its basis in a nonconceptual consciousness of phenomena-in-themselves. For unless it does, it is difficult to see how concept-laden observation of phenomena could be distinguished from the concepts or ideas endemic to Hume's sense impressions or Kant's raw material of experience (that have no articulable relation to an external reality).

An external reality *qua* phenomena-in-themselves also have extensiveness.[17] In this sense physicists are indisputably conscious of phenomena-in-themselves as occupying a lived region of space. It is not only the moving bodies that are explicated kinematically in classical mechanics of which they are conscious as volumetric or dimensional. It is the fringe patterns and diffractions of light in quantum mechanics as well.

This volumetric or dimensional feature of phenomena-in-themselves, in given regions or axes, is presupposed by the objective or approximately objective measurements of physicists. Moreover, as Compton notes,[18] such physicists would presuppose extensiveness for objectively viewing phenomena from different perspectives and for insisting that an empirical-theoretical construct prove essential in several independent experimental contexts.

It is in the context of experimental setups that physicists realize an ideal aim of manipulating phenomena. And this evidences a connection between manipulation and features of phenomena-in-themselves, *e.g.* extensiveness, that invoke practical insights related to existentialism and phenomenology. For phenomenal manipulation is unintelligible apart from both the volumetric 'stuff' of the world and the purposes of physicists. Thus physicists may manipulate phenomena in electrodynamics for purposes of telecommunication or the elements of fusion for heat. But their purposes of manipulation are phenomenologically related to their incontrovertible consciousness of themselves as freely choosing persons.

This is an important point because freely chosen purposes reflect a connection between phenomenology and 'existential action.' This is especially evident in applied physics such as that of engineering. Samuel C. Florman asserts in his generally disregarded *The Existential Pleasures of Engineering* (1976), for instance, that 'they [engineers] enjoy the neat way in which this physical world is subject to manipulation.'[19]

Given such manipulation, what is the significance of Heidegger's sarcasm concerning 'science wishing to know nothing of Nothing'[20]? Heidegger's notion of Nothing (*what is not*) *is often understood by existentialists as being juxtaposed to the indifferent way the world is (what is*) to which physicists inauthentically and passively submit. But Florman notes an existentially moral significance of modern engineering that has existential import:

> The main goal has always been to understand the *stuff* of the universe, to consider problems based on human needs..., to propose solutions, to test and select the best solution, and to follow through on a finished project. *Existential* delight has been the reward every step of the way - for the observer, the user, and particularly for the doer.[21]

Thus contrary to Heidegger, or to the misinterpretation of his thought, the submission by physicists and engineers to *what is* is a submission to an objective world in order to manipulate it for their purposes.

The notion of such purposes, in virtue of being phenomenologically experienced together with freedom, comprises the very refutation of the Kantian thesis that freedom - unlike determinism - does not have a basis in concrete human experience. I shall address the alleged conflict between determinism and freedom shortly. I now note that the freely chosen manipulation of phenomena is concomitant with the fourth feature of phenomena- in-themselves. For the latter could not be manipulated independently of their having *continuity*.

Continuity is their changing or evolving in an orderly and causally predictive manner. The manner of such predictable change has arguably had a better metaphysical articulation in Nagarjuna's early Eastern philosophy, in which a cause is neither identical with nor distinct from an effect,[22] than Parmenides' pre-Socratic explication of an impossible transformation of what is (*Being*) into what is not (*Non-Being or Nothing*). For the Parmenidesian rejection of *Nothing* anticipated the myopic positivistic notion that the meaning of a word is *something* for which it must stand (as criticized by Ludwig Wittgenstein, in his *Philosophical Investigations* [1][23]). But it also anticipated the logocentric appeal to reason in modern metaphysics. This metaphysics, in reflecting Parmenides' exhaustive appeal to reason, ignored the fact that causal determinism - to which Hume and Kant gave critical expression - is not merely based on reason. Rather it is based paramountly on the physicist's phenomenological experience of phenomena- in- themselves.

Observational consciousness of phenomena-in-themselves is inextricably a part of the physicist's observation of changing phenomena, say the oscillating brightness of a glowing tube screen or motion of celestial bodies. For observation of such bodies or oscillating glows, whether instrument aided or not, involves their changing and yet retaining identities.[24] This is precisely what permits a given phenomenon to be causally explained.

Such explanation, while less epistemologically secure than indubitable Cartesian ideas, is more ontologically anchored than the Humean sense impressions from which the empiricist idea of causality is derived. Whereas the idea of Humean causality did not knowingly correspond to or 'mirror' an external reality (*pace* Richard Rorty's *Philosophy and the Mirror of Nature* [1990]), the phenomenological experience of continuity is the non-ideational basis for ascribing reality of phenomenal regularities. *Pari passu* it is the basis for positing the reality of the theoretical entities, processes, and properties in terms of which such regularities are understood.

Such an understanding is conceptually and phenomenologically linked to the perduring applicability of theoretical laws that, construed in terms of a logico-mathematical schema, imply various predictions.

Phenomenological experience and predictability

Let me integrate phenomenological experience into a schema endemic to predictability after reiterating an existential freedom and purpose of physicists.

For physicists who predict must resolutely confront the brute fact of an external reality that *is how it is* independently of what they will, wish or think *it is*.

Thus if existentialism is associated with maximizing our 'own unfettered freedom'[25] (Dostoevsky) or with realizing 'the indifference of *being*'[26] (Sartre) or with the 'marvel of all marvels: that what is - *is*'[27] (Heidegger), then physicists are existential exemplars. For they acknowledge the world's 'coefficient of adversity' (Sartre) in order to predict and manipulate it. Its manipulation and prediction are achieved to a degree that is virtually unobtainable in any other human endeavor.

Simply stated, this endeavor may be related to a logico-mathematical schema. Thus an observational description (o_1), say of a rolling ball in classical mechanics, may be linked to a theoretical description (t_1), say of a particle having mass m_0 and velocity v_0 at time t_0 on a given axis (wherein $o_1 \approx t_1$).[28] By the same token o_1 might describe a light source aimed at a medium with inhomogeneities (holes in screens, edges of bodies etc.) and be linked to t_1 in terms of a microphysical particle M having wave properties, that moves in region Δ_x on an axis.[29]

But it is entirely ignored that o_1, either in classical or quantum mechanics, will also involve an observational consciousness of both phenomena and experimental apparatus as in themselves having *ceteris paribus* the features noted above. Hence the symbol 'o_1' may be modified as 'o_{1c}' to denote such consciousness as well as to ontologically link concept-laden notions of phenomena to phenomena-in-themselves. For besides phenomena-in-themselves providing an extra-conceptual dimension to observational entities denoted by o_{1c}, the nonconceptual consciousness of such entities involves the physicist's implicit awareness of his purposive activity.

The immediate purpose of this activity might be to coordinate the theoretical description t_1 with a theoretical law (L_o) of momentum $p = mv$ in classical mechanics or $\Delta_p \approx h/\Delta_x$ in quantum mechanics. For these laws together with the rules of logical deducibility (L) will yield a theoretical description *qua* prediction in terms of the schema $t_1 \wedge L_o \; L \; t_2$.[30] This represents the theoretical element of the schema wherein t_2 is itself coordinated with an observational description ($t_2 \approx o_{2c}$) in order to derive or deduce an observational prediction.

Notwithstanding the distinction between 'theoretical deduction' and 'theoretical derivation' wherein derivation by deduction belies that 'identification [of perceptual descriptions with theoretical ones] does not occur within the theory,'[31] a general theory in terms of which a law is understood is more than a Kantian idealization of empirical reality or an 'instrumental device.'[32] Thus a theoretical prediction (t_2) in classical mechanics may posit definite values characterizing a particle's coordinate at each moment of time and t_2 in quantum mechanics may posit a value only up to accuracy Δ_p of the uncertainty in the momentum. But the truth or approximate truth of what Δ_p or momentum p specifies in t_2 is no less understood in terms of o_{2c} than the truth or approximate truth of o_{1c} is understood in terms of t_1.

This renders an ontological significance to t_1 and a theory in terms of which it is understood that permits the ascription of truth or limited truth to the theory from which o_{2c} is derived. o_{2c} is coordinated with observational phenomena, whether a fringe pattern on a screen or rolling ball in observable regions. And the latter are inexorably connected with phenomena *qua* phenomena-in-themselves being independent of will, wish, or thought, their having individuality, their occupying a lived spatio-temporal region, and their changing in ways that are successfully predicted.

It is such predictability that raises troublesome predicaments for logocentric Anglo-American philosophies of science which ignore existential phenomenology. For although phenomenology links successful prediction with the theoretical truth of L_o, such philosophies are inhibited by the fact that truth is not logically implied by success. Bas C. van Fraassen, in *The Scientific Image* (1980), asserts that 'empirical adequacy does not require truth; in my view, science aims only at empirical adequacy and anything beyond that is not relevant to its success.'[33] And van Fraassen's assertion is complemented by Kurt Hubner's *Critique of Scientific Reason* (1985) wherein 'since, according to the rules of logic, true statements can follow from false ones, we can now state... that nature has not expressly said no to the content of the [successful] theory - but neither has it said yes to it.'[34]

The predicament is, that independently of truth, there can be no intelligible talk of scientific *knowledge!* Phenomenology, while not embracing the knowledge of physics as paradigmatic knowledge, posits reality of observational referents and hence truth or approximate truth to the laws and theories in terms of which such referents are predicted and manipulated. It is interesting in this respect that Nancy Cartwright and Ian Hacking lend their thought to the notion that truth is an *existential ingredient* in causal explanations.

Hacking argues that experimentalists do not believe in electrons because they 'save the phenomenon' but rather because they use them to manipulate and create new phenomena.[35] And Cartwright backs up this argument by adding that causal and theoretical explanations have been conflated: Although truth is an 'external characteristic' of laws, the physicist must be committed to the real *existence* of the cause if he is to accept a given causal account.

But to this it can be added that to accept the causal account together with the reality of the cause is to accept the truth of the laws and theory in terms of which the cause and account are understood. Why is it then that many Anglo-American philosophers understand theories as being empirically successful as opposed to true? Besides being unable to ameliorate the Humean-Kantian bifurcation of reality from theory (wherein $o_{1c} \approx t_1$ and $t_2 \approx o_{2c}$), they have another related epistemic dilemma.

This dilemma may be specified in terms of the 'underdetermination-of-theory-by-data' thesis (UTD)[36] wherein any observational data or phenomena may be in principle explicated and predicted by empirically equivalent but logically inconsistent theories or theoretical constructs. Thus in addition to being unable to correlate such constructs with observation independently of observation-theoretical concepts in question, they cannot (will not) accept such possible inconsistency. It is not a violation of their cherished principle of noncontradiction [~ (p∧ ~ p)] to hold that two or more inconsistent theories are empirically adequate or successful. But it is an ostensive violation to hold that they are all true.

This dilemma concerning truth is not burdensome to phenomenology in the same manner by virtue of its appeal to an 'ambiguity' of pre-scientific experience. Such experience is alluded to in Sartre's existential phenomenology when he says that 'It [*being*] always overflows whatever knowledge we have of it - just as it is presupposed by all our questions and by consciousness itself.'[37] And such experience becomes evident in Merleau-Ponty's notion of an 'ambiguous life in which the forms of transcendence have their *Ursprung*, and which, through a fundamental contradiction, puts me in communication with them, and on this basis makes knowledge possible.'[38]

Thus it becomes possible to say that phenomena-in- themselves, notwithstanding their general features, 'overflow' or 'transcend'any univocal observational description (o_{1c}) or theoretical description (t_1) to which the former is reduced. This is wholly consistent with the notion that t_1, being axiomatic and typically involving the application of integral and differential calculus, reduce the inexact empirical predicates of o_{1c} to exact ones relevant to a given theory. Thus the phenomenological understanding of theory underscores several points: that inconsistent theoretical constructs may attach to the same hypothesis, say to de Broglie's hypothesis concerning the particle-wave duality of light (λ = h/mv = h/p, where λ = wavelength, h = Planck's constant, p = mv = magnitude of the momentum of a moving particle); and that Aristotle's influential principle of noncontradiction is applicable to *statements* about reality and not to *reality per se*.

Who can say a priori what *reality* can or cannot be? We may ask this question without questioning the notion either that no two propositions of a physical system should be contradictory or that from no two propositions, say p and q, should it be possible to deduce two inconsistent propositions, say r and ~ r. Thus although ~ R may express the denial of the particle-wave hypothesis and should not be deducible from a physical system that also implies R, there is no reason prima facie why R - the particle-wave hypothesis - cannot otherwise be a proper implication or part of the system. Moreover the ambiguity of the system's hypothesis would reflect, at a theoretical level, an ambiguity at the experiential level.

It is noteworthy that phenomenological experience and inconsistent theoretical constructs do not imply either a relativism criticized by Newton-Smith or a 'post-rationalist' instrumentalism rejected by Karl Popper. Popper overlooks the fact that we do not demand of a photon that it abide by the principle of noncontradiction but rather of our statements about it that they abide by the principle. Thus on the one hand Popper asserts:

> There is no reason whatever to assume... that the single photon is in the two different and incompatible states at one time; any more than to assume that it moves through both slits at one time. These are different possibilities, different virtual states, inherent in the experiment; and these different possibilities are weighed and measured by the propensities.[39]

One need not deny Popper's notion of propensities, wherein de Broglie's pilot waves can 'be best interpreted as waves of propensities,'[40] in order to reject his notion that the denial of such propensities is linked to a 'general anti-rationalist atmosphere.'

That he makes this link to such an atmosphere is, on the other hand, clear when he states (three lines below the above longer quote):

> The general anti-rationalist atmosphere which has become a major menace of our time, and which to combat is the duty of every thinker who cares for the traditions of our civilization, has led to a most serious deterioration of the standards of scientific discussion.[41]

Scientific discussion is enhanced by the notion of phenomenological ambiguity if for no other reason than that it reminds philosophers that reality need not conform to paradigmatic standards of rationality, viz. Parmenides', Aristotle's, Kant's or anyone else's.

Ironically, Popper recognizes this elsewhere when he concedes that 'It is our Aristotelian and essentialist habit of expressing ourselves... which leads to the subjective interpretation of possibilities, and which makes it difficult to accept an objective propensity interpretation.'[42] But such an interpretation exacerbates an essentialist habit by returning the subjective interpretation to an objective Aristotelian one in which reality conforms a priori to a rationalist interpretation.

Phenomenology draws attention to the fact that there may be different but veritable theoretical interpretations some of which attach to the same hypothesis as in de Broglie's case. Besides there being no reason for supposing that theoretical interpretations cannot comprise a growth of increasingly truer theories (similar to Popper's 'verisimilitude'), there is no reason for supposing that phenomena-in-themselves change from culture to culture. It is not cultures that comprise truth- conditions for scientific truth-claims but rather phenomena-in-themselves of which scientists are observationally conscious.

In closing it is important to reiterate that consciousness also yields a phenomenological basis for a deterministic metaphysics. Such a metaphysics is not a vacuous conceptual-linguistic construction without a point of contact with reality, in phenomenology. But in the absence of phenomenology it easily peregrinates into the Kantian problem of scientific truth-claims presupposing a truth-valueless (*synthetic a priori*) principle of causal determinism. One obvious strategy of neo-positivists is to reject the notion that modern physics presupposes causal determinism.

Thus, for example, it is often held that Newton's inverse square ($F_{gr} = G\, m_1 m_2/r^2$) made no immediate reference to a cause and that equations of quantum mechanics embrace an indeterminism by virtue of being probabilistic. But the equations of quantum mechanics are *deterministic* of probabilities and the *application* of Newton's laws do inescapably suppose regularities of mass particles or bodies as well as causes, *e.g.* in terms of gravity.

Thus neo-(post-) positivist philosophers either do or do not acknowledge a metaphysical principle of causal determinism. If they do not, the application of scientific theories to empirical reality is unintelligible. If they do, scientific truth-claims are reliant prima facie on a truth-valueless metaphysical principle. The embracement of such a principle, besides generating the latter dilemma, has pejoratively influenced scientific epistemologies: Kurt Hubner asserts that 'the very occurrence and rise of the sciences... must be considered something determined by a historical situation;'[43] Thomas Kuhn holds that scientific paradigms (disciplinary matrixes) condition scientists to view phenomena certain ways;[44] Paul Feyerabend argues that scientists are 'caused to accept or reject' uninterpreted observation sentences in response to sensory stimuli;[45] and Peter Gardenfors' recent article, 'Induction, Conceptual Space, and AI' (*Philosophy of Science* [1990]), explicates human reason through evolution.[46]

Existential phenomenology rejects such appeals to evolution, conditioning, or strict determinism without relinquishing a basis for them in the physicist's phenomenological experience. The experience of the continuity of phenomena-in-themselves is the basis for a deterministic metaphysics that, while not a metaphysics of which empirical truth can be ascribed, does veritably reflect how physicists are conscious of nature and not of themselves.[47] If they were themselves exhaustively characterized by a continuity endemic to phenomena-in-themselves, then they together with their truth-claims would be exhaustively such phenomena. It is not such phenomena of which truth is ascribed but rather of statements and theories about it. And it is on pain of conceptual and phenomenological absurdity that physicists are not incontrovertibly aware of their freedom to formulate or not to formulate such statements and theories.

Notes

1. Harvey E. White, Ph.D., Sc.D., *Modern College Physics*, 6th Edition (D. Van Nostrand Co., Inc., OP 1981), p. 564.

2. Alan Chalmers, *Science and Its Fabrication* (Minneapolis: University of Minnesota Press, 1990), p. 59.

3. Cf. *Existentialism From Dostoevsky to Sartre*, Ed. by W. Kaufmann (New York: New American Library, 1975), p. 13, and *The Confessions of St. Augustine,* Tr. by J. Ryan (New York: Doubleday, 1962), p. 342. Kaufmann links Dostoevsky's emphasis on 'man's inner life... and his decisions' to St. Augustine. And St. Augustine's assertion *inter alia* that 'I will to be and to know' as well as his peppered references to human consciousness (in response to skeptics) link him to an existential epistemology.

4. Chalmers, p. 15.

5. W.H. Newton-Smith, *The Rationality of Science* (London: Routledge & Kegan Paul, 1981), pp. 27-28. Notwithstanding my criticism, it is my opinion that Newton-Smith's work has made the best case *ceteris paribus* for a realism and rationality in physics than any book in the last decade.

6. Ibid., p. 28.

7. Ibid., p. 28.

8. Ibid., p. 35.

9. Ibid., p. 43.

10. *The Structure of Scientific Theories*, Ed. by Frederick Suppe (Chicago: University of Illinois Press, 1977), pp. 126-127.

11. John Compton, 'Natural Science and the Experience of Nature,' from *Phenomenology in America*, Ed. by James Edie (Chicago: Quadrangle Books, 1967), pp. 80-95. Also see Compton's excellent analysis in 'Phenomenology and the Philosophy of Nature,' *Man And World* 21, 1988, 65-89.

12. Compton, 'Natural Science and the Experience of Nature,' p. 90.

13. Cf. Tom Bethell's 'A Challenge to Einstein,' *Nat Rev* XLII, 1990, 60.

14. Cf. P. Ubertini, A. Bazzanu, R. Sood, R. Staubert, T.J. Sumner, and G. Frye, 'Hard X-Ray Spectrum of Supernova 1987A on DaY 407,' *The Astrophysical Journal* 337, 1989, L19, and G. Chui's 'Pulsar Vanishes: Astronomers' Baffled,' Knight News Service, July 23, 1989, for this quote (my emphasis).

15. Compton, 'Natural Science and the Experience of Nature,' p. 90.

16. Ibid., p. 90.

17 Ibid., p. 90.

18 Ibid., pp. 90-91.

19 Samuel C. Florman, *The Existential Pleasures of Engineering* (New York: St. Martin's Press, 1976), p. 94.

20 Martin Heidegger, 'What is Metaphysics?,' Tr. by R.F. Hull and A. Crick, from *Existence and Being* (Chicago: Henry Regnery Co., 1968), p. 328.

21 Florman, p. 113 (my emphasis).

22 Cf. Robert Trundle's and R. Puligandla's 'Beyond the Linguistic and Conceptual: A Comparison of Albert Camus and Nagarjuna,' *Darshana International* XVI, 1976, 1-12, and their book Beyond Absurdity: *The Philosophy of Albert Camus* (Lanham, Maryland: The University Press of America, 1986).

23 Notwithstanding Wittgenstein's quote from St. Augustine's *Confessions* (I.8.), this picture of language anticipates the synthetic propositions of Logical Positivism wherein such propositions have a factual meaning in virtue of being based upon the empirical objects to which they refer. Cf. Ludwig Wittgenstein's *Philosophical Investigations*, Tr. by G.E.M. Anscombe, 3rd Edition (New York: The Macmillan Co., 1971), p. 2e.

24 Compton, 'Natural Science and the Experience of Nature,' p. 90.

25 Fyodor M. Dostoevsky, *Notes From Underground*, Tr. by C. Garnett (New York: The Macmillan Co., 1923). From Kaufmann, p. 71.

26 Jean-Paul Sartre, *Being and Nothingness*, Tr. by Hazel Barnes (New York: Philosophical Library, 1956), p. 192.

27 Martin Heidegger, p. 355.

28 Cf. Stephan Korner's excellent analysis of an 'identification'of empirical state descriptions with theoretical ones in *Experience and Theory* (London: Routledge & Kegan Paul, 1969), pp. 183-184. Also see N.R. Hanson's 'Is there a Logic of Scientific Discovery,' *Readings in the Philosophy of Science*, Ed. by B.A. Brody and R.E. Grandy (Englewood Cliffs, New Jersey: Prentice-Hall, Inc., 1989), pp. 398-409. Hanson argues that reasoning *towards* hypotheses may be as legitimate an area for inquiry as reasoning *from* them.

29 Cf. B.M. Yavorsky and Uy A. Seleznev, *Physics: A REfresher Course*, Tr. by G. Leib (Moscow: MIR Publishers, 1979), pp. 420, 486-487, or D. Halliday and R. Resnick *Physics: Part II* (New York: John Wiley & Sons, 1986), pp. 1210-1212. Subsequent reference to equations of quantum mechanics may be made to these pages.

30 Korner, pp. 188-189.

31 Ibid., p. 169.

32 Sartre, p. 445.

33 Bas C. van Fraassen, *The Scientific Image* (Oxford: Clarendon Press, 1980), p. 198.

34 Kurt Hubner, *Critique of Scientific Reason*, Tr. by P.R. Dixon, Jr., and H.M. Dixon (Chicago: The University of Chicago Press, 1985), p. 140.

35 See Ian Hacking's 'Experimentation and Scientific Realism,' *Phil Topics* 13, 1982, 71-88, and Nancy Cartwright's 'When Explanation Leads to Inference,' *Phil Topics* 13, 1982, 111-121.

36 See, for example, John Worrall's 'Scientific Realism and Scientific Change,' *Phil Q* 32, 1982, p. 223, for a classic articulation of the UTD thesis. Also see D. Goldstick's and B. O'Neill's 'Truer,' *Phil Sci 55*, 1988, 583-597, for pragmatic considerations about conflicting theories.

37 Sartre, p. xx.

38 M. Merleau-Ponty, *Phenomenology of Perception*, Tr. by Colin Smith (London: Routledge & Kegan Paul, 1978), p. 365.

39 Karl R. Popper, *Quantum Theory and the Schism in Physics*. From the *Postscript to the Logic of Scientific Discovery*, Ed. by W.W. Bartley, III (Totowa, New Jersey: Rowman and Littlefield, 1982), p. 156.

40 Ibid., p. 142.

41 Ibid., p. 156.

42 Ibid., p. 128.

43 Hubner, p. 124.

44 See Thomas Kuhn's 'Second Thoughts on Paradigms.' From Suppe, pp. 459-482. Kuhn compares the student of science developing a 'learned relationship, an acquired perception of analogy' (p. 472) to Johnny 'being programmed to recognize what his prospective community already knows' (p. 475).

45 See Paul Feyerabend's 'An Attempt at a Realistic Interpretation of Experience,' *Proceedings of The Aristotelian Society*, New Ser., 58, 1958, 143-170. Suppe notes (p. 637) that, for Feyerabend, 'The observation language for a class C of observers will be, roughly, those sentences which members of C are caused to accept or reject in response to sensory phenomena' (Suppe's emphasis).

46 See Peter Gardenfors' 'Induction, Conceptual Spaces, and AI,' *Phil Sci* 57, 1990, p. 92.

47 There can be an Aristotelian or Thomistic twist to phenomenology wherein nature's lawfully describable (predictable) behavior reflects an Unmoved Mover's thinking or God's immanent intelligence. Phenomenology, notwithstanding Sartre's atheism, might underscore that the intelligibility of such intelligence proceeds *pari passu* with an Unmoved Mover or God being implicitly conscious of their intelligent behavior. A Sartrian objection that such a theological or quasi- theological conception of God *qua* reality is contradictory might be held to reflect the failure to consider that the principle of noncontradiction is applicable to statements about

reality and not to reality per se. The further objection that we could not discursively grasp such a reality and that it is thus a meaningless notion might invoke the further response that the notion of mind (consciousness) and body (or Sartre's 'pour-en-soi' and 'etre-en-soi') are no less paradoxical if fruitful for articulating human experience.

6 The evolution of science and the 'tree of knowledge'

Jean Marie Trouvé

Introduction

Trevor J. Pinch and Weibe E. Bijker have advocated the need for an integrated approach towards the study of science and technology[1]. Although they observe differences between science and technology such as the larger 'number of social groups relevant to the technology case', 'the more heterogenous environment in which technology develops', they 'take such differences to be a matter of degree, and not as representing any fundamental distinction as to how the two cultures should be approached'[2].

What the work here demonstrates is precisely that to account for such differences is more than a matter of degree. What is needed, in order to cope with such differences, is a kind of approach to specialised knowledge very different to traditional approaches in cognitive studies of science, as the approach propounded and the theory too briefly developed here shows[3].

On the applicability of scientific knowledge models to technology

The argument raised against the possibility of applying to technology some of the models elaborated in cognitive studies of science goes, in fact, a step further. The application is bound to the capacity of the model to take into account, in particular, a greater variety of groups, more groups, a wider context extended to the society itself, a more heterogenous environment, the multidisciplinarity of technological knowledge. In short, the model must be capable of accounting for more groups and more contents, or in other words for a greater flexibility of knowledge. But that objective, as the discussion will show, cannot be fulfilled with the kind of approach to scientific knowledge proposed by Pinch and Bijker nor by the improvement of existing models of scientific knowledge offered in cognitive studies of science. The reason is that they have already failed either to realise or to take into account the full flexibility of scientific knowledge itself, although they themselves have contributed with their observations and their work to increase the degree of flexibility of scientific knowledge unveiled so far. Since models of scientific knowledge have already failed to account for the flexibility of scientific knowledge observed in studies of scientific knowledge taken as a whole, why should they be suited for taking into account the flexibility of technological knowledge?

The heart of the objection can be summarised as follows. The image of scientific knowledge resulting from the models of scientific knowledge developed in cognitive studies of science underestimates the flexibility of scientific knowledge. These are inadequate to describe technology, and in particular technological progress, not really because the application is unjustified on the basis of insuperable differences between science and technology but because they are inadequate to fully describe scientific knowledge and its development in coping with the high degree of flexibility of specialised knowledge, either scientific or technological.

Before presenting the approach we follow, in order to cope with the flexibility of scientific knowledge, with the resulting definitions, descriptions and representations of its development - including the structure, the growth and the evolution of scientific knowledge - let us briefly examine why so many underestimate different approaches and therefore cannot cope with its flexibility.

Coping with a global approach to the flexibility of scientific knowledge

Historians of science have amply shown that scientific knowledge varies over time. Sociologists of science have observed that for a given problem and a given time, scientific knowledge varies between groups of scientists. Laboratory studies have shown that scientific content varies during the process of production. In scientific controversies claims vary during the negotiation process. Scientific knowledge even varies from situation to situation for the same scientist. All these variations in scientists' discourse define the flexibility of scientific knowledge. The notion accounts for the fact that scientific knowledge is open to more that one interpretation. It is also referred to as 'the interpretive flexibility of scientific knowledge'[4] or 'the participants' interpretative work'[5].

Whereas these observations taken as a whole unveil the great flexibility of scientific knowledge, taken in the context of each kind of approach, they lead to an image of scientific knowledge which does not benefit from an integrated approach[6]. The resulting image given by each approach underestimates the flexibility of scientific knowledge. The corresponding models of development of scientific knowledge owe their existence and their coherence to the fact that they are 'underloaded' with flexibility. They were never tested with an 'extra load' coming, for instance, from one of the challenging approaches[7]. This is a challenge that none of the models, with the corresponding approach to scientific knowledge, can afford! Their validity is at stake!

Furthermore, the need to devote time and work on one locus of observation introduces another limit to the range of actors and contents taken into account in the analysis[8], increasing the underestimation of the flexibility of scientific knowledge, while, at the same time, backing the conservative tendencies in science via the study of the most visible actors and contents. Thus the analyst who studies scientific knowledge at the research front, where it is being produced or (and) legitimised, can be tempted to consider and take into account only the kind of flexibility he is able to observe and know at first hand. He will then describe it as 'the interpretative flexibility of scientific findings'[9], introducing a restrictive definition of flexibility which imputes flexibility only to findings and new knowledge.

The reasons for opposing an integrated approach to the flexibility of scientific knowledge cannot be disentangled from opposition to the study of science as a whole, as one topic. They are bound to heavy trends in research practices and traditions, often imported from the humanities, and social and human sciences. The strong empirical tradition has encouraged the multiplication of case studies whilst the open, consensual rejection of theorising[10] has discouraged the kind of work which could have contributed to unify existing and crumbling knowledge of scientific knowledge. Can one more empirical case study improve our understanding of the development of scientific knowledge?[11] Cognitive studies of science should pay a particular attention to theorising because after philosophy and history of science, they are, in their turn, solicited to provide models of scientific knowledge and of knowledge systems to other specialities not only within the whole area of science and technology studies but also to external disciplines. Unfortunately, they do not fall short of the same kind of criticisms as those addressed to any other disciplines[12]. Alluding to the 'strong programme', Steve Woolgar notes: 'it is used to denote a preoccupation with low level theoretical concerns'[13].

But it would be unfair to limit criticisms to science and technology studies, whereas the models of science and technology come, for a large part, with the methods, approaches and *a priori* conceptions about the nature of science and technology, from the social and human sciences and the humanities. Without enquiring into *a priori* conceptions about the nature of scientific knowledge within those disciplines[14], it will be difficult to know how these *a priori* conceptions have drawn their way through science studies and if the image of scientific knowledge, as it emerges from science studies, is the result of the study of scientific knowledge or if it is largely predetermined. We are loosing 'any benefit to be gained from the study of the specific area of science', Steve Woolgar remarks and pleads: 'Perhaps it is now time to try and make science talk to sociology rather than the other way round'[15]. Is the work done in cognitive studies of science assessed from criteria specific to science studies or from the predominant criteria of evaluation used in traditional disciplines? Lack of autonomy of cognitive studies of science and dependence from traditional disciplines do not facilitate their progress!

Coping with the learning process and hierarchized contents

A different sort of flexibility is required. It is most important for two kinds of people: the analyst of scientific knowledge who has to cope with the technicality of scientific knowledge, and the future scientist who, in schools and universities, has to become familiar with scientific knowledge. Although the flexibility of scientific knowledge can be observed by anyone, the analyst does not experience it in his attempt to cope with the technicality of scientific knowledge. What he experiences, as any science student does, is something very different, it is quite another matter: it is the quasi absolute rigidity of scientific knowledge. He is not familiar enough with scientific knowledge to express an opinion on it, to propose a different or improved version, or to oppose one content to another. He remains dependent on the content he is taught or has been told about. The analyst is here entangled in two kinds of contradiction between:

1. the necessity of accounting for scientific knowledge flexibility in order to cope with scientific knowledge change and the extreme rigidity he experiences because of his lack of familiarity with the technical content of scientific knowledge;

2. the objective 'to study the very content of scientific knowledge' and the huge cost to fulfil it, in terms of effort, work, training and time needed to acquire the technical competency generally 'demanded' in any science training centre.

A science student perceives a given subject differently as understanding of it (supposedly) improves during his school life. In the learning process, the flexibility is given by the different contents which become one after the other the science student's perception of a given subject. And there is no reason to expect the learning phenomenon to stop in the laboratory after university training; nor once he is a well known scientist. The learning process is such that it leads to the decrease of the rigidity of scientific knowledge. It renders scientific knowledge more flexible. It is essentially an individually experienced phenomenon. How can this kind of flexibility be made compatible with the other one, the only one to be taken into account in cognitive studies of science? Is there a continuity between knowledge learning and knowledge production?

This kind of flexibility is very different from that currently described. It does indeed show that there exist non-equivalent forms of knowledge, or in other words 'hierarchized' forms of knowledge. Far from training centres and their regular public, learning processes can be expected to be always at work and non-equivalent forms of scientific knowledge (or contents) can be expected to emerge even if drawing a demarcation line between the two is problematical.

Any theory of scientific knowledge must take into account hierarchized forms of knowledge. The approach has then to be evaluative in some way just because hierarchically ordered forms of scientific knowledge already exist. Ultimately the theory must account for both kinds of contents: non-hierarchized and hierarchized contents.

Does scientific culture exist?

Many scholars justify the application of methods, concepts, types of analysis and approaches employed in specialities dealing with cultural phenomena on the grounds that scientific culture is one culture amongst other cultures: 'The treatment of scientific culture as a social construction implies that there is nothing epistemologically special about the nature of scientific knowledge: it is merely one in a whole series of knowledge cultures (including, for instance, the knowledge systems pertaining to 'primitive tribes')[16]. Consequently, the anthropological method is perfectly suited for laboratory studies in particular: 'En anthropologie en effet, il est usuel de se familiariser intimement avec une culture que l'on ne partage pas. En appliquant la meme vision generale aux sciences, on peut produire une analyse fine des details de la production scientifique sans etre oblige de croire ou de partager dans l'explication les prejuges et les notions des chercheurs observes'[17].

This kind of epistemological justification bears the following message with a strong political content:[18]

1. Science, as a topic, should only be an area of application of methods developed in traditional disciplines.

2. In order to cope with the very content of scientific knowledge, being trained in traditional disciplines and informed on the class of objects they deal with is more important and relevant (priced) than being informed on science and trained in science studies: in other words, ignoring science, science contents, science studies, should be highly recommended.

3. Consequently, the only experts capable of dealing with science matters should come from the stock of existing disciplines.

4. The stock of disciplines should never grow[19].

Can it be said, from what we know of culture and science, that anything like a scientific culture exists, as far as scientific knowledge is concerned? People who have popularized science as well as those who have studied the phenomenon of the popularization of science strongly doubt that anything like a scientific culture exists. The main reason invoked is that growth of knowledge through specialization renders impossible any building up of a minimal knowledge background, which could be communicated and shared by anyone[20]. Even from the work of the same scholars, this obviously cannot be said of scientific knowledge because of its flexibility, the multiplicity of the groups of scientists, the very small size of most of the groups of scientists, the short life of scientific claims in some situations or because, as we have seen in the last section, of a hierarchy between certain contents which can be invoked to invalidate one given content by another.

As far as scientific knowledge is concerned, scientific culture has to be defined by reference to scientific knowledge. Will the attempt result in anything which deserves the name of culture? In any case, the idea of scientific culture is in need of clarification. The pretentions of the cultural approach to deal with scientific knowledge are unjustified. What did scholars mean by culture in the first place?

Is the paradigmatic approach of any use to the study of scientific knowledge?

As far as scientific knowledge is concerned, any direct use of Thomas Kuhn's work is of little help. The American historian was indeed more interested in the description of the development of science as a whole than by the development of scientific knowledge. The main criticisms have to be addressed to his followers who, in their attempts to develop the study of scientific knowledge, did not choose the right piece of Kuhn's work to imitate.

Kuhn has explained the kind of difficulties he faced when trying to describe normal science periods with the concept of consensus - a necessary step in order to give sense to his idea of revolutionary change. He only found pronounced disagreement on scientific definitions: 'the concept I had been seeking did not exist'[21]. He finally retained, as an equivalent concept, what the scientific profession was taught and accepted: the standard scientific problems and their solutions. The choice of the concept of paradigm should be viewed as accompanying i) the broadening of Kuhn's interest as a result of his efforts to account for normal science periods in order to save his idea of revolutionary change ii) and the shift of the focus of his research away from specific scientific knowledge matters.

The most significant result in Kuhn's work, as far as scientific knowledge is concerned, is the lack of consensus on scientific definitions. But since Kuhn did not develop his theory from this result, he left future analysts of scientific knowledge with the task of coping with the flexibility of scientific knowledge and describing its development. Unfortunately, most of his followers, in cognitive studies of science, inspired by the concept of paradigm (in its multirevisited sense!) did not build upon the right result, concept and piece of work. They failed to realise that the concept of paradigm, in its original meaning, does not refer to what the concept of consensus applies and refers to[22]. In Kuhn's work, in Kuhn's own words, normal science periods are characterised both by a paradigm, and a lack of consensus on scientific knowledge (even on definitions)[23].

This is why, even if Kuhn's description of scientific development is retained - the alternance of normal science periods with revolutionary science periods - the transfer of the pattern of cognitive studies of science (no matter how this occurred!) with controversies replacing revolutions and consensuses replacing normal science periods must be rejected. Building the study of scientific controversies upon a consensus between controversies instead of a dissensus inevitably leads to an image of scientific knowledge where the flexibility is, from the start, underestimated. Furthermore, the existence of the model then imposes constraints on the use of the empirical observations themselves to make them fit the model. It encourages analysts to ignore the diversity of the contents and to underestimate the importance of the differences between contents in order to save the coherence of the model, even if their own observations tend to contradict their presuppositions. This is why one can observe that in order to avoid the overheating of the model due to a too great flexibility, a cooling system is often surreptitiously re-introduced to annihilate part of the flexibility[24].

It is still rather a mystery, to say the least, that the discovery from which Thomas Kuhn has developed his work and his description of scientific development - the importance of solved problems for scientists - has been almost ignored. Any theory of scientific knowledge, seeking to describe the development of scientific knowledge, should, at least, deal with this discovery. Empirical studies should either confirm or invalidate the discovery, and any theory be built upon the result. Or, a theory of the development of scientific knowledge should be built independently of the issue, as we did with the theory presented here, and an answer should come from the theory itself. As it was shown in the last section, the empirical result of Thomas Kuhn - the importance of solved problems in science - is here discovered and confirmed by the theory as one of the main results.

Coping with the constraints on the flexibility of scientific knowledge

So far we have stressed that the task undertaken copes with the flexibility of scientific knowledge, an objective which converges with that advocated by Michael Mulkay. It thus diverges from that of almost all approaches in cognitive studies of science: to study the very content of scientific knowledge.

One should not see here a shift from this early and main objective. On the contrary, it is rather the realization that the development of scientific knowledge can be fully described and understood only if all contents are taken into account in the analysis and not only a few of them as this inevitably occurs when the analyst deals directly with technical contents: indeed, 'analysts are found into making narrow selections from the full range of discourse available in principle'[25]. This is why analysts of the content of scientific knowledge, as far as their final ambition is to propose a general description of the development of scientific knowledge, should not only ignore the diversity of accounts but also and above all ignore content[26]. Consequently, coping with the flexibility of scientific knowledge and considering the phenomenon of flexibility as a research topic in its own right are proposed as the approach capable of overcoming the kind of difficulties encountered in dealing with the very content of scientific knowledge.

The following approach will unveil and describe precisely what is lacking and hidden in 'local approaches', that is the knowledge of the place of the few actors and of the few contents analyzed on one research spot within the global distribution of knowledge and of actors involved in the study of the same, or related problem, but generally scattered in different locations. The resulting picture, in the literal sense since the distribution emerges from the theory in the form of representations or figures, shows the kind of constraints that the global distribution of knowledge puts on the degree of flexibility of new knowledge and on the number of acceptable alternatives[27].

The difficulty is: how to make a description of the development of scientific knowledge emerge from the analysis of the flexibility of scientific knowledge? The basic problem is what to do with the flexibility of scientific knowledge?[28]

The flexibility and the unity of scientific knowledge

What are we to make of Kuhn's observation concerning the lack of consensus on scientific contents during normal science periods? We are left with an atomised knowledge, with a multiplicity and a diversity of contents and with the multiplicity and the diversity of the groups of scientists sharing those contents. We are left with the task of explaining not only the existence of two kinds of traditionally opposing features of science but also their articulation: on one side lack of consensus, incommensurability, revolutions, ruptures, discontinuities, the multiplicity and the diversity of contents and of groups of scientists, the smallness of the groups involved in the production of scientific knowledge, local knowledge, and on the other side large areas of consensus, solutions to problems, evolution, the continuity and the unity of knowledge, universal knowledge, the collective dimension of knowledge.

And what are we left to do from the state of the art in most of studies of scientific knowledge, that is from a much too consensual description of scientific knowledge? We are left with the task of introducing a dissensus when traditional approaches describe and account for a consensus, that is with the task of introducing more groups, more contents, a far greater diversity and heterogeneity. We have then to introduce smaller groups and atomised knowledge where shared knowledge is described by others. We have thus to look for a smaller unit of analysis of knowledge and of groups than those in use[29].

When introducing more contents, we also introduce more information, technicality, complexity and richer details, that is all the richness and technicality of scientific knowledge. It means that in using a 'horizontal approach' to the study of scientific knowledge - 'going larger' through scientific knowledge in introducing many contents - we realise what 'vertical approaches' to scientific knowledge attempt to do - 'going deeper' into scientific knowledge.

Using a smaller unit of analysis of knowledge and of consensus means that observations and analyses should be made at a structural level microscopic[30] as regards to the structural level at which consensuses are dealt with. And as far as unity of knowledge is concerned, it should be described at a macroscopic structural level as regards to the level at which atomised knowledge is described. This second level is expected to be itself macroscopic as regards to the structural level at which consensuses are traditionally studied. In any case, two linked levels of structure should emerge from the theory whereas current approaches deal with only one level.

Describing atomised and unified knowledge is to deal with the static dimension. The dynamic dimension should be revealed in the description of the atomisation and unification processes. But both processes should not be considered as leading from one feature to another as if the two features were autonomous. The theory is perfectly capable of accounting for the dependency of one feature on the other. Indeed, the order of realization of the two objectives - at first, to account for atomized knowledge, since the task set up is to cope with the flexibility of scientific knowledge, then to account for unified knowledge - suggests the way of defining the unity of knowledge. It can be defined from atomized knowledge. The task left then is to show how the different contents forming atomized knowledge are linked to each other. Unity of knowledge gains thus its meaning from the overall set of links where all contents are held together.

Atomization of knowledge is then expected to describe an increase of flexibility and unification of knowledge a decrease of flexibility. That is also to say that atomised and unified knowledge have to be described as two relative concepts, accounting for the fact that atomised knowledge is always characterized by some degree of unity and that unified knowledge is always characterized by some degree of atomicity.

Atomization and unification of knowledge can be presented as a function of the flexibility of scientific knowledge. Both phenomena thus refer to one process: the variation of the flexibility of scientific knowledge. In other words, it is the flexibility of the flexibility of scientific knowledge rather than the flexibility alone which must be used in order to cope with the development of scientific knowledge.

This is coherent with the fact that since they are supposed to characterize scientific knowledge, atomicity (or dispersion) and unity exist simultaneously. They are not and cannot be treated as exclusive features as this occurs when they are depicted successively. At any time, in any situation, for any problem, scientific knowledge is characterized by a certain degree of atomicity and unity. Even atomised, and because flexibility is a major characteristic of scientific knowledge which allows us to depict scientific knowledge as more or less atomized or more or less unified, scientific knowledge should be expected neither to have been deprived of its propriety of interconnectedness nor to have lost contact with the rest of knowledge.

Consequently, if a certain degree of atomicity and of unity is a constitutive feature of scientific knowledge and if both components exist simultaneously, both processes of atomization and of unification should develop in an ahistorical dimension. Although the theory deals perfectly with the historical dimension of scientific knowledge and accounts for development over time, it also shows, with the case of these two processes, the way to introduce a dynamic process which evolves in an ahistorical dimension[31].

The growth of knowledge and the learning process

Linking knowledge production process with the knowledge learning process

In cognitive studies of science research fronts are defined as places where knowledge is produced and negotiated. The definition is equivocal and limitative: it reduces the process of negotiation to the phenomenon of production of knowledge. Are noisy places known for being exclusive and exceptional innovative centres? Furthermore, the geographical definition of research fronts does not fit with the main objective of cognitive studies of science: to study the very content of scientific knowledge. Research fronts could be expected to be defined in cognitive terms too.

But the main criticism is of the silence covering the way analysts have reached research fronts. The total lack of information - which is compatible with the fact that reaching geographical places is not a real source of difficulties with the availability of planes, trains, cars[32] - suggests that research fronts are reached unproblematically, instantaneously. Indeed, analysts of scientific knowledge use research fronts as the points of departure of their work. But how have they managed to raise themselves, most of them with no background in science - what difference does it make to have some training in science? - in the few months or weeks or days or hours devoted to an empirical study of the very content of scientific knowledge, to learn and to assimilate all the knowledge that scientists they study had to assimilate during twenty years or so of schooling and researching in order to be, one day, in the position to produce knowledge, that is precisely to reach the research front[33]? Too many scholars, through their own way of dealing with scientific knowledge, propagate the damaging idea that assimilation and transfer of knowledge might be fulfilled by a sort of passive, infectious or telepathic processes where learning, schooling, researching, working, effort and time are totally absent[34].

Analysts of scientific knowledge fail to acknowledge and take into account the universality of learning. This partly explains their naivety and ease with which they impute to scientists an extra-expertise: the capacity of inventing interpretations and alternatives almost at will. They fail to realize that in coping with the description of the way they improved their own understanding of scientific knowledge they might have obtained a model for describing the progression of scientists towards research fronts[35]. But in reaching research fronts instantaneously, or in letting one think that one can reach them thus, they are left with not much to say not only about their 'transport system' but also about the scientists' 'transport system'. In the end they remain silent about the (old) knowledge to which the new knowledge, the knowledge being produced (that that analysts observe), owes its existence and to which it has to be linked. It is from silence over the access to research fronts that most analysts are given the kind of uprooted knowledge they need to justify their arguable approach to scientific knowledge.

The research front is here considered as a problematical notion. Instead of being the point of departure of the approach followed, it is, on the contrary, the point of arrival. Describing the ways to reach the research front is to describe the learning process, that is to account for progress in knowledge, thus to describe the successive, richer, forms of knowledge that the learner acquires one after the other before being capable of producing knowledge. If we fulfil this task, not only will we know how to reach the research front but also we will have linked knowledge production to knowledge learning, new knowledge to old knowledge. We will have shown the continuity and the interconnectedness of knowledge, and discovered the existence of a similarity between two processes: knowledge production and knowledge learning.

The description of the learning process is interesting for another reason: it will be used to define the growth of knowledge. Indeed, one way of observing the growth of knowledge is to be part of the phenomenon itself so as to be in the ideal position for describing new successive scientific contents as they appear, just as in the learning process.

The structure of scientific consensus

The notion of consensus is treated as a problematical one. In science studies it refers to a widespread agreement. But it is of no interest to say that many agree if nothing is known about exactly who, how many, to what degree they agree on exactly what. The notion of consensus gains its meaning from the notion of knowledge in the link which unites two categories to each other to form one single concept with two components: one made of knowledge, the other made of the people sharing the content. The analyst of scientific knowledge should therefore be vigilant and guided by an insatiable curiosity for finding out over which contents, statements, ideas, and truths any consensus is supposed to be realized.

Inquiring into and unveiling the finer structure of a consensus would be informative. Would the consensus be made by a hard core of specialists who would be the most active and creative, whilst the majority only followed the leaders? Should a consensus be noteworthy just because of its hard core? If it owed its importance to its size rather than its hard core, it would gain in quantity what it loses in quality.

The resulting poor informative value emerging from the way consensuses are dealt with in studies of scientific controversies both creates and distorts the controversy/consensus distinction. Lack of precision with which a consensus is known brings into question its very reality. This precision is not sharp enough to reveal a clear distinction between consensus and dissensus. This means that analysts of controversies do not have an adequate concept to state that, when and if, controversies collapse into consensuses. With any increase in the precision with which a consensus is known, you may have to explain why controversies also collapse into dissensuses, why dissensuses do not always result in controversies, why large portions of consensus exist simultaneously with controversies, endangering the model upon which this kind of study is built, the scope and the validity of studies of scientific controversies.

Finally, the notion of scientific knowledge is itself treated as a problematical notion. Scientific knowledge is localized nowhere. There is not one place, where knowledge is concentrated like any industrial product, where anyone could go and take charge of it. What should be known of a given subject is only known to us in the form of references to the subject. These come from highly specialized journals, school books and popular magazines, private correspondence and oral material, interviews, conferences, new and old material: the only criterion retained in the choice of a reference is its existence. Knowledge is concentrated neither in one nor in a set of references. It is unequally distributed in all the references to a given subject, in equivalent, non-equivalent and hierarchized references. The concept of 'reference to scientific knowledge' is thus substituted for the notion of 'scientific knowledge'.

The problem of access to the knowledge of a subject is that of access to the collective dimension of that knowledge from and through a series of successive accesses to multiple individualized references. The widespread consensus is not sufficient to give scientific knowledge the collective dimension from which it gains its meaning. The overall distribution of knowledge among the references must be known. The place of each individualized reference must be known within the general distribution. The analyst has to determine how the collective dimension of scientific knowledge can emerge from the individual dimension of scientific knowledge, and how all the references are interconnected?

Let us choose a debated subject (indifferently settled or still going on) with the group of people who make theirs one (or *the*) interpretation. If it is the largest group it is traditionally considered as the consensus. For the analyst, difficulties arise in determining the size and the composition of the consensus (who belongs to the consensus?) and these continue when he tries to determine the content of the interpretation that members of the consensus are supposed to share. It then appears that there is just no content specifically attached to the group of people designated by the consensus. There is no 'referenced knowledge' available.

The need to know what the content of the supposed consensual interpretation consists of leads the analyst to seek the referenced knowledge attached to each member of the consensus. The analyst discovers not only a great variety of contents with a multiplicity of 'grains of consensus' but also the existence of non-equivalent, or comparable, contents[36]. Some introduce and deal with all sorts of information and comments while others do not, and remain silent. Some deal with empirical observations as well as with theoretical arguments, express a well argumented and nourished opinion, while others do not relate their opinion with others' opinion nor with the related experiments, concepts or theories. This last sort of opinion, in its lowest form, looks like an uprooted form of knowledge, of a dependent and non-usable nature, a pure affirmation, an inspired rather than an illuminating form of knowledge.

The absence of information and comments in certain references as compared with their presence in others is very informative not so much as to the different ways of understanding the subject than as to the differing ways people experience the rigidity of the subject. We can thus try to order the references according to the relative degree of rigidity of the subject felt by people. We must expect to find with the highest degree of rigidity the lowest forms of knowledge at the bottom of the distribution of the references and with the lowest degree of rigidity the highest forms of knowledge at the top. In going from lower to higher forms of knowledge, one would observe the decrease in the rigidity of the subject while one would improve one's understanding of the subject as in a learning process.

In order to account for the distribution of the references, we do not cope directly with the references one by one. We start with the study of a given consensus. First of all, let us uncover and formalize the phenomenon which accounts for our previous observations of the references. It is used to describe and represent the structure of scientific consensus and, with it, the distribution of the references.

We interpret the absence of information, data and comments in some references (but found in others) as the absence of reaction and response from these references when solicited for an answer. Silences are thus interpreted dynamically and used to hierarchize references. The phenomenon can be described at the level of the consensus as follows. When the consensus must face a new piece of information it is split up into new 'units of consensus', each made of the group of individuals or 'unit of individuals' sharing a new enriched content or 'unit of knowledge'. While new units of consensus appear, some individuals, present in the original consensus, do not appear in the new units of consensus. With the references they are part of, they no longer participate in the process of the enrichment of knowledge and no longer take part in the progress of our understanding of the subject studied. The references concerned have exhausted their knowledge resources. Any of the new units of consensus is itself solicited by other information found in the remaining references, and the process is continued until all the references are exhausted.

The phenomenon can be characterized, and summarized, by the following twofold effect:

1. A multiplying effect: the production of new consensuses, from an original one, associated with new enriched units of knowledge;

2. A reducing effect: the decrease in the number of individuals present in the new consensuses in regards to the number of individuals present in the original consensus.

The 'Tree of Consensuses' thus obtained defines the structure of scientific consensus. Units of consensus of the same generation are considered as equivalent whereas successive generations of units of consensus are hierarchically ordered. Any unit of consensus is connected to any other unit of consensus. The interconnection is total. Three examples of connection are drawn in figure 1. For each generation, the number of units of consensus is indicated according to the hypothesis that any unit of consensus splits up into two new units. Other representations of the structure of scientific consensus are given in figures 2, 3 and 4.

The structure and the growth of scientific knowledge

Since the unit of knowledge is one component of unit of consensus, the structure of scientific consensus represents the structure of scientific knowledge. It also represents the distribution of knowledge per unit of consensus, and indirectly, per individual and per reference.

The structure uncovered so far accounts for atomized knowledge and for the atomization process. It is the configuration of divergence, also represented in figure 2. We will now demonstrate the existence of a convergent structure and of a unification process leading either to a few units of knowledge, regarded as the research front, or to a unique unit of knowledge, regarded as the solution of the problem.

Let us introduce a little quantitative data. The first unit of consensus, or rank 1 consensus, from which all others emerge, is made of all the individuals present in the references to the problem considered. Its size, measured by the number of people (it is a finite number) is the biggest of all the units of consensus of the tree of consensuses. Indeed, when the rank 1 unit of consensus is split up into new units of consensus, or rank 2 units of consensus, some of the individuals present in the rank 1 unit of consensus do not appear in the new units. This means that the number of people found in the set of rank 2 units of consensus is smaller than that found in the rank 1 unit of consensus. We begin again with the rank 2 units of consensus. Each rank 2 unit of consensus is split up into rank 3 units of consensus. The number of people present in the set of rank 3 units of consensus is smaller than that of people in the set of rank 2 units of consensus. In other words, when a rank is added, the number of consensuses increases and the number of individuals decreases. It means that there comes a rank where the number of consensuses cannot increase due to lack of people. The extreme case occurs when the number of consensuses equals the number of people. The top rank units of consensuses cannot be split up since there are no knowledge resources left to solicit them, and their number can no longer increase. Either the resulting structure accounts for the state of the art of the problem or, before the extreme case is reached, with the addition of another rank another process, following the atomization process, must start and account for another phenomenon.

The only possibility left with the regular decrease in the number of people is the grouping of people in lesser units of consensus. The first rank to be added then stops the atomization process and starts, with the decrease in the number of units of consensus, the unification process. We are in a configuration of convergence, represented in figures 3 and 4.

The configuration also defines the growth of knowledge. Indeed, successive units of knowledge show a better understanding of the problem. As successive ranks are crossed knowledge of the problem increases and improves until the research front, or the solution, is reached. This accounts for a process of knowledge growth and is used to define the growth of knowledge. The process of knowledge learning has thus been identified with a true process of knowledge growth. The univocal relation between knowledge growth and knowledge production has been broken, allowing for the accounting for continuity between knowledge learning and knowledge production, old knowledge and new knowledge.

The concept of line of growth has been introduced. It is designed to account for two facts: 1) in order to reach the research front, or the solution, one does not have to progress through the whole structure. Following one line of growth suffices; 2) several lines of growth lead to the research front, or to the solution.

Two distinct parts of the configuration are shown: the field of divergence and the field of convergence. Two important differences between the two are to be mentioned. The first concerns the average contribution to the growth of knowledge per unit of consensus: it decreases in the field of divergence and increases in the field of convergence, as a rank is added. The second concerns the average contribution to the growth of knowledge per individual: it is greater in the field of convergence than it is in the field of divergence since the number of individuals present in any of the ranks of the field of convergence is smaller than that in any of the ranks in the field of divergence. It reflects the far greater efficiency and competencies of individuals present in the field of convergence as knowledge producers.

The evolution of scientific knowledge

Each configuration is attached to one problem, one subject or one branch of science upon which knowledge has grown. Any theory of scientific knowledge with universal ambitions must thus deal with all the problems scientific activity is composed of. It must deal with all the branches of science. The task consists of integrating all of them, first to link two of them, then a third to the first two and so on to cover the entire field of scientific knowledge.

In order to deal with the evolution of scientific knowledge not only must the theory account for the way that all the branches of science can fit together but also how new branches of science can be integrated into older ones. The approach we follow allows us to treat these two objectives as one: indeed, we will show why and how one branch of science emerges from another.

The content of a given unit of knowledge reflects the opinion and the understanding of a subject shared by the corresponding unit of individuals. Accordingly, it is different, in various degrees, to the content of any other unit of knowledge. This characteristic reflects the tendency of units of consensus to pull the tree of consensuses out of the limits of the related subject to which it is attached and confer to units of consensuses the propriety of evolving independently in their own interests.

One unit of knowledge can reveal a content from which a new subject in need of explanation emerges. The corresponding unit of consensus then becomes a candidate for generating a new tree of consensuses attached to the new subject.

The reproduction mechanism is shown in figure 5. It operates in an open system, which is coherent with the fact that we are dealing with evolution. More precisely, to go from a given configuration to the next, we have to go from the generating unit of consensus to the rank 1 unit of consensus of the next generation. As indicated earlier, the rank 1 unit of consensus is the biggest unit of consensus since it contains all the individuals present in the configuration. Consequently, the generating unit of consensus is, generally, not big enough to generate by itself the growth of a new configuration. A complementary mechanism has to be introduced to ensure the reproduction process hampered by the exhaustion of human resources: we have termed this 'loading'.

Furthermore, the references attached to the generating unit of consensus do not exhaust the new subject. The missing reference, out of the scope of the generating configuration, must be reached and the new subject 'loaded' with them.

From one subject to another, from one branch of science to another, a growing area of scientific knowledge is covered so that the entire field can be covered. The representation of the evolution of scientific knowledge is given in figure 6. The concept of 'line of evolution' has been introduced. Lines of evolution link successive generations of configurations to each other[37].

The importance of solved problems in the development of scientific knowledge

The most remarkable feature of the tree of knowledge is that it does not diverge much, contrary to what the predominance of the fields of divergence over the fields of convergence would lead to expect[38]. How can one explain this characteristic?

Let us look for an explanation in the resulting effects of the two kinds of configurations on the shape of the tree of knowledge. The more numerous and wider the configurations of divergence are and the wider the fields of divergence, the wider is the tree of knowledge. But an increase of width does not automatically result in an increase of divergence. The widening also has to occur in proportion to the increase in height, that is in proportion to the corresponding growth of knowledge.

There are indeed factors opposing the tendency to increase the divergence of any configuration, which control the widening of the evolution front, that is the atomization of knowledge phenomenon. At first, there is the fact that any divergence also produces knowledge and thus contributes to the heightening of the evolution front. Although a strong divergence is very unproductive, the resulting effect on the growth of knowledge is never nil, so that the regular heightening of the evolution front with the knowledge growth goes against the divergence effect.

There is another limiting factor: any divergence uses up many units of consensus. This means that the rapid exhaustion of human resources consecutive to any increase of the divergence imposes an insuperable limit to the increase of divergence.

As for the effect of the field of convergence, it annuls a part of the effect of divergence, or all of it as in the case of a complete convergence. Any new convergence, while lifting up the evolution front, corrects the divergence effect and re-centres the tree of knowledge. The cumulative effect of convergence would control the cumulative effects of divergence.

This characteristic thus reflects the importance of the complete convergence configurations in the tree of knowledge, an importance which goes far beyond their actual location. It emphasizes the role of solved problems and well-known branches of science over the general development of scientific knowledge, a result which rediscovers and confirms Kuhn's crucial observation: the importance of 'problem solutions that the profession has come to accept as paradigms'[39].

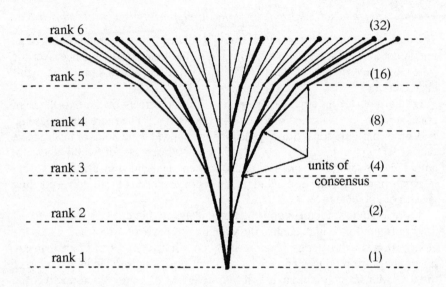

figure 1

The Structure of Scientific Consensus
or
The Tree of Consensus
(configuration of divergence)
The Atomization Process
(increasing divergence)

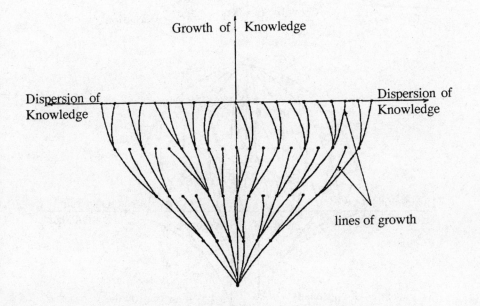

figure 2

The Structure and the Growth
of Scientific Knowledge

Configuration of Divergence
(decreasing divergence)

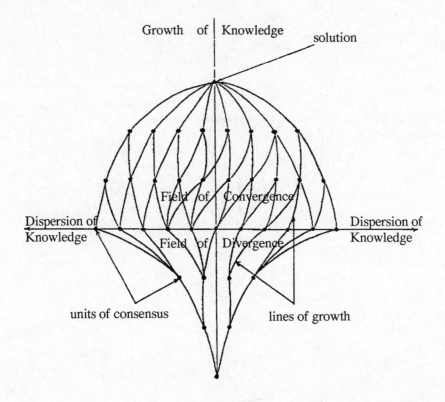

figure 3

The Structure and the Growth
of Scientific Knowledge

Configuration of Convergence
(closed configuration)
(complete convergence)
(increasing convergence)

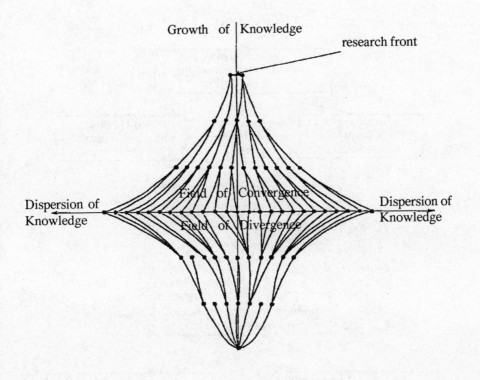

figure 4

The Structure of the Growth
of Scientific Knowledge

Configuration of Convergence
(open configuration)
(incomplete convergence)
(decreasing convergence)

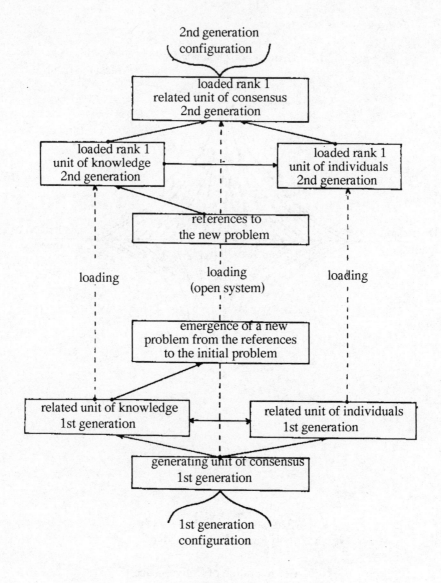

figure 5

The Reproduction Mechanism
of Scientific Knowledge

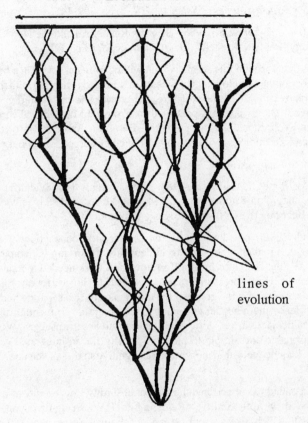

figure 6

The Evolution of Scientific Knowledge

or

The Tree of Knowledge

Notes

1. Pinch T.J. and Bijker W.E., 1984, 'The Social Construction of Facts and Artefacts: or How the Sociology of Science and the Sociology of Technology might Benefit Each Other', *Social Studies of Science,* XIV, 3, pp. 399-441.
2. Pinch T.J. and Bijker W.E., 1986, 'Science, Relativism and the New Sociology of Technology: Reply to Russell', *Social Studies of Science,* XVI, 2, pp. 347-60.
3. The ideas developed in this article, which will be fully described in a book, are part of A Theory of the Development of Scientific Knowledge dealing with the definition and the representation of the structure, the growth and the evolution of scientific knowledge. They were first presented in Trouvé J.M., 1986, 'The Structure of Scientific Evolutions', *Senses of Science,* European Association for the Study of Science and Technology IVth Meeting, Council of Europe, Strasbourg.
4. Op. cit. note 1, pp. 431.
5. Mulkay M., Potter J., Yearley S., 1983, 'Why an Analysis of Scientific Discourse is Needed', chap. 7 in Knorr-Cetina K., Mulkay M. (eds.), 1983, *Science Observed,* London-Beverley Hills- New Dehli, Sage Publications.
6. Pinch and Bijker, pleading for 'a more integrated conception of science and technology', impute to the 'contingent demands of carrying out empirical work in these areas' (the cost in terms of effort, time and work needed to gain expertise in highly technical literature is such that analysts tend to carry on research in the domain where this expertise is the greatest) the emergence of two separate bodies of work, one on science the other on technology. But they fail to mention that the same contingent demands are at work within science studies themselves. Would it not be more reasonable and simple to plead in priority for an integrated conception of scientific knowledge within the domain of cognitive studies of science? Op. cit. note 1, pp. 432.
7. Gad Freudenthal has underlined the kind of absurdity constructivism leads to when fully loaded with flexibility. Referring to P.W. Bridgman's operationalism, according to which 'every scientific concept should be associated (...) with a single operation', he notes that: 'In just the same way constructivism would lead to an unlimited proliferation of scientific products'. Freudenthal G., 1984, 'The Role of Shared Knowledge in Science: the Failure of the Constructivist Programme in the Sociology of Science', *Social Studies of Science,* XIV, 2, pp. 292.
8. M. Mulkay, J. Potter and S. Yearley review a series of approaches to scientific knowledge for which they uncover the different ways the analysts fail to consider the diversity of the contents. Op. cit. note 5.
9. The expression is used by Pinch and Bijker. Op. cit. note 1, pp. 409.

10. Knorr-Cetina K. and Mulkay M., 1983 in Knorr-Cetina *et al* (eds.), op. cit. note 5, pp.13.

11. Empiricism can also be used as an alibi for justifying almost any kind of policy and work resulting, as in the case of the French 'Science Technologie et Societe' programme (1978-1983) to the absolute crumbling of the area, to shoddy work and to the discouragement of most people who could not see any progress. Strong encouragements towards more 'd'etudes factuelles, d'enquetes sur le terrain, d'etudes empiriques et statistiques' (Hamelin N., 1983, *Pandore,* 22, Paris, Maison des Sciences de l'Homme, pp.51-3) had been given. 'Il faut travailler beaucoup sur des objets locaux. Nous n'en sommes pas a l'heure de la grande synthese. Des etudes de cas, des objets locaux, multipliez et croissez!' (Hamelin N., 1983, *Pandore,* 24, pp. 66- 8). This advice creates the illusion that on one hand work on the terrain is a fairly good substitute for reflexion and for specialization to research in science studies and on the other hand time for reflexion and specialization will always come early enough! The dramatic issue was later described: 'The area of STS has become, not what the founders of Pandore wished it to be, but exactly the 'magma' that they wanted to escape from.' (1984, *EASST Newsletter,* III, 1, pp. 17). This is an illustration of the disastrous effects caused by a certain kind of empirical work when it is not deeply rooted in a strong research tradition.

12. In almost any area of science and technology studies, some people have linked the poverty of results and, sometimes, of entire specialities to the poverty of theorizing. It is particulary true of technology studies which widely 'do draw on models from other disciplines': Johnston R., 1984, 'Controlling Technology: An Issue for the Social Studies of Science', *Social Studies of Science,* XIV, 1, pp. 103. The need for more concern with questions of theory and method in science policy was desperately expressed by Karl Kreilkamp: 'Anybody familiar with the field knows how little attention is being paid to these questions', 1973, 'Towards a Theory of Science Policy', *Science Studies,* III, pp. 3.

13. Woolgar S., 1981, 'Interests and Explanation in the Social Studies of Science', *Social Studies of Science,* XI, 2, pp. 368.

14. In order to uncover *a priori* conceptions about the nature of scientific knowledge, the author has explored the following line of research: to study the very content of scientific knowledge (quantitative methods, physics models, equations systems, concepts, bits and pieces of natural and biological theories, and other references to scientific knowledge) once borrowed, as it is applied in humanities and social and human sciences; Trouve J.- M., 1981, *Essai Critique sur les References Mystificatrices au Savoir Scientifique,* monogr., Paris, Centre Science Technologie et Societe, Conservatoire National des Arts et Metiers.

15. Woolgar S., 1983, 'Irony in the Social Study of Science', chap. 9 in Knorr-Cetina K. *et al* (eds.), op. cit. note 5, pp. 262.

16. Pinch T.J. and Bijker W.E., 1984, op. cit. note 1, pp. 401-2. Scientists are considered by most scholars as engaged in 'the production of scientific culture'; Knorr-Cetina K. and Mulkay M., 1983, pp. 9 in Knorr-Cetina K. *et al* (eds.), op. cit. note 5. 'The

meta-analyst rejoices in the restoration of science to its 'proper' place as a cultural activity'; Chubin D.E. and Restivo S., 1983, 'The 'Mooting' of Science Studies: Research Programmes and Science Policy', chap. 3 in Knorr-Cetina K. *et al* (eds.), ib., pp. 72.

17. Latour B., 1982, *La Science Telle qu'elle se Fait,* Paris, Pandore.

18. Even if the message is not endorsed by all the authors of such justifications, it refers to a reality and a status quo that others have pleaded for and maintain.

19. The conception of an area of research devoted to the study of science and technology considered only as an extension of existing disciplines bringing new financial resources was expressed by the people in charge of the French STS programme (see note 11): 'travaillez dans le STS mais ne reniez pas vos origines! (...) Le STS n'est ni une discipline, ni une ecole de pensee ni un programme minimal: STS est un niveau d'analyse' (Hamelin N., 1983, *Pandore,* 24, pp. 66- 8). It should not come as a suprise then that five years later the study of science and technology has not developed yet into a speciality, still less into a discipline. It just is not in the interest of the stock of disciplines that a new challenging discipline should emerge.

20. Blanc J. (Directeur de la cite des sciences et de l'industrie, Paris), Hulin M. (Directeur du Palais de la Decouverte, Paris), 1987, 'Information Scientifique et Societe', *Ecole d'Ete de Science de l'Information,* Poitiers, France.

21. Kuhn T., 1977, *The Essential Tension,* Chicago, The University of Chicago Press.

22. Thomas Kuhn was not of great help in clarifying this aspect. Not only new meanings were added to the original meaning of paradigm, but he kept on using, as he explains himself, the notion of consensus instead of paradigm to describe normal science: "consensus' rather than 'paradigm' remains the primary term there used when discussing normal science'; ib. Preface. If Thomas Kuhn's readers have 'contents' in mind, they thus should understand dissensus when they read consensus!

23. The confusion resulted in the misunderstanding of what the notion of incommensurability was applied to in Kuhn's work. 'What in Kuhn's theory was an incommensurability between paradigms has, in the constructivist sociology of scientific knowledge, become an incommensurability between individuals and the things they manufacture.'; Freudenthal G., 1984, op. cit. note 7.

24. See Freudenthal G., ib.

25. Mulkay M., Potter J., Yearley S., 1983 in Knorr-Cetina K. *et al* (eds.), op. cit. note 5.

26. See Russell S., 1986, 'The Social Construction of Artefacts: A Response to Pinch and Bijker', *Social Studies of Science,* XVI, 2, pp. 331-46.

27. 'Local approaches' take advantage of the fact that isolation, in masking the existence of non-local knowledge and of the great number of different phenomena and data which are linked to the 'local' ones, leads to ignore how, if and to what extent the overall network (or the interconnectedness of science) constrains and structures the

scientist's context and interpretation. See Gringas Y. and Schweber S.S., 1986, 'Constraints on Construction', *Social Studies of Science,* XVI, 2, pp. 372-83.

28. The author makes his the two ways of formulating the kind of problem to be tackled offered by Gad Freudenthal: 1) 'the very same reasons that make it possible to talk of one length which may be differently measured, also require that we speak of one solution x, prepared through different selections.'; 2) 'to discover descriptive means of allowing one to construe identity in non-identity'; Freudenthal G., 1984, op. cit. note 7.

29. The size of the unit of analysis is an important quality since it defines the precision with which a consensus is known. Only the precision can warrant, in the study of scientific controversies, that, when and if, controversies collapse into a consensus. Lack of precision marks the fact that controversies may have also collapsed into a dissensus.

30. The size of the unit of analysis (which defines the structural level at which observations are made) is an important feature in so far as it largely determines the change and the rate of change in science that the theory is capable of explaining. This is why, in order to explain the rapid rate of change observed in science, the paradigm community associated with the speciality is favourised against the paradigm community associated with the discipline. See Hoch P.K., 1987, 'Institutional Migrations and New Scientific Specialities', *EASST Newsletter,* VI, 1, pp. 11-5.

31. This is only one in a whole series of issues dealing with the antinomic aspects to be overcome in order to account for the structure, the growth and the evolution of scientific knowledge in one coherent approach to the development of scientific knowledge.

32. 'Reading' should be added to the 'transport system' used to reach research fronts less geographically-bound like controversies, which can be considered as being a move away from a geographical definition of the research front.

33. Who controls, in cognitive studies of science, what the learnt knowledge looks like? One way of dealing with the problem would be that analysts accept the duplication of others' case- studies.

34. This is the kind of absence which makes one think that cognitive studies of science have never been so far from science policy matters at a time when more and more scholars are looking for science policy implications from social studies of science results.

35. A model accounting for progression towards research fronts independently of the contents and of the people involved is reflexive, like the one we propose. It applies to scientists as well as to analysts of scientific knowledge.

36. For a detailed analysis of consensus, see Trouvé J.-M., 1982, 'La Structure Fine d'un Consensus Scientifique', *Fundamenta Scientiae,* III, 3/4, pp. 279-96; Trouvé J.-M., 1985, 'A Microanalysis of the Structure of Scientific Consensus', *Explorations in Knowledge,* II, 2, pp. 1-17.

37. The theory is of an evolutionary type, although it was not developed from the perspective of evolutionary epistemologies.
38. This is an example illustrating that the observations made at one structural level of analysis are not extendable to another structural level of analysis. What is observed for the growth of knowledge is not observed for the evolution of knowledge. One phenomenon concerns the level of the configuration, the other the level of a multitude of generations of configurations.
39. Kuhn T., 1977, op. cit. note 21.

7 A sociological perspective on disease

Kevin White

Introduction

This paper identifies three approaches to the concept of disease in the theoretical literature in this field. These approaches, the empiricist, the normative and the social are shown to be based on three major positions in epistemology. These are respectively, the Cartesian, the Hegelian and the Nietzschean. It is shown, that despite apparent differences the Cartesian and Hegelian positions share the fundamental premise that knowledge of disease is ultimately possible. Thus while the empiricists believe disease to exist in an ahistoric and absolute way, the Hegelians are equally convinced, that stripped of contaminating social influences, the true nature of disease will be revealed. By contrast the position argued for in this paper, the social or Nietzschean, is that diseases are constituted by the society in which they are located. That is to say, they are social phenomena without residue; they do not have an independent existence. This position, while at first sight difficult to sustain, is supported by its ability to resolve conceptual and theoretical problems in the theory of disease held to by Cartesians and Hegelians.

Three concepts of disease

Disease has been conceptualised by sociologists and philosophers in three basic ways. These may be labelled the empiricist, the normative and the social. For the empiricist disease is defined unproblematically as that set of phenomena which presents itself as a fact within the context of a natural science methodology.[1,2] In this perspective disease is a purely physiological occurrence, defined essentially as the absence of health.[2] This approach is most vulnerable to criticism, given the recent developments in the history and philosophy of science. As Hesse[3] and Feyerabend[4,5] have argued, 'facts' do not present themselves prior to theory formation. They are indeed constituted by the theory. Any claim that nature presents itself unproblematically to science is logically problematic.

The second approach, the normative, overlaps with the first. It attempts to argue that disease categories are problematic because they are used to explain the wrong types of phenomena. What are really social problems - drug addiction, gender identity and so on - are wrongly called 'diseases'. However, when researchers who adopt this position attempt to specify what it is they are discussing, they are left with a vacuum. Disease becomes something which is indefinable[6] or the characteristic only of singular individuals.[7] Theories of disease which take as their focus the individual are in essence based on a seventeenth and eighteenth-century contractarian theory of society. They assume that society is the sum of individual wills rather than that the form and content of those will derive from the organisation of society.[8] Such an approach does not facilitate a theory of disease from a sociological perspective.

The third approach, the social, as developed in this paper, argues that disease categories are the outcome of the society in which they exist. There is no pursuit of the ontology of disease but rather an attempt to locate disease phenomena in its historical and cultural specificity.

This classification of approaches to the question 'what is a disease?' reflects the major philosophical positions (following[9]; see also[10]) in answer to the question 'how do we have knowledge of the world?' Those who adopt the empiricist position are philosophically Cartesians. Language is a series of signs that reflects a pre-existent reality. Since that reality is fixed there must be a fixed method for gaining an understanding of it. This is reflected, it is argued, in the development of the physical sciences, whose theories and methods are correspondences to that reality.

Normative theories of disease on the other hand are Hegelian. They take for granted that rationality is historically and sociologically specific. Our knowledge of the world is always developing, becoming more inclusive, eliminating errors and more closely approximating the truth. The present is better than the past, not so much because it is nearer the truth, but because it represents the historical process of synthesis and progress. However, the Hegelians still wish to preserve the idea that there is a truth. Given enough time, the Hegelians argue, real disease entities or conditions will be fully illuminated while mistakes, like calling what are really social actions - homosexuality for example - diseases, will be transcended.

The position argued for in this paper, the social, is Nietzschean. It rejects that which unites the Cartesian and Hegelian position: the assumption, explicit in the first, implicit in the second, that there is an ultimate truth. Rather, there are socially and culturally specific understandings of a socially and culturally constructed reality. One does not exist - independently or prior to the other - but mutually constitute each other.

Let us consider the Nietzschean critique of Hegelian approaches to disease. In the Middle Ages, medicine had to be congruent with the dominant ideas of law and theology. It was 'not simply a thing in itself but the expression of a particular view of reality'.[11:385] In this period the professions of law, medicine and theology were institutionalised remedies for the fall - which was also the source of disease. It was only following the English Revolution that an attempt to rationally understand the occurrence of epidemics could develop. In particular it allowed conceptually for the development of the idea that there was a link between urban life and illness, and further that there was a relationship between urban living conditions, illness and immorality. 'Filth bred moral filth, disease bred moral disease'.[12:268] Within this context debates between the miasmatists and contagionists took place. Tesh[13] argues that miasmatic theory was accepted not because it was 'proved' but because its main competitor, contagionism and the associated quarantine program it specified for disease prevention, was politically unacceptable. Tesh however confuses the issue by retaining a residual notion of 'science' as a non- economic and non-political phenomenon. Her analysis cogently illustrates that what counts as scientific is the outcome of social, political and economic struggle. However, the Hegelian epistemology informing her position leads to the contradictory position of suggesting that 'anti-contagionists began to confuse the scientific with the economic and to use economic arguments to advocate a scientific position'. [13:328]

The Nietzschean position, when left at a philosophical level is open to criticism, not only of relativism (a Cartesian criticism) but, ironically, of ahistoricism (a Hegelian criticism). For example, in the work of M. Foucault[14], the leading representative of Nietzschean epistemology, it is possible to demonstrate 'that the relationship between categories and objects is purely conventional and arbitrary'.[15] In this context it is then easy to drift from sociology back into philosophy: ideas take on an existence independently of the social groups who formulate them. It is only if knowledge claims are located inside the social context that gives rise to them - that society is seen as a reality sui generis - that 'truth' can be grounded. If societies and knowledge formation are conceptualised as long term processes[16] then a sociological approach to knowledge can be sustained. For example if Foucault's corpus is seen to be a documentation of the long term process of rationalisation in Western culture, then his investigations can be sociologically locked back into the dynamics of a real society.[15] In this way concerns about the ultimate 'truth' are undermined since 'the linkages between categories and phenomena are ultimately guaranteed by the logic of human arrangements'.[15:200]

The question 'what is disease?' has been addressed from a variety of perspectives. Biological determinism holds that disease is the absence of health and that manifestations of it are recognisable solely by natural science methodology.[1,17] This approach adopts a staight-forward empiricist epistemology which is premised on a value-free and normatively neutral concept of facts.[1,17] This position has been persuasively attacked, by Hesse[3] and Feyerabend[4,5] among others.

Similarly, Mario Bunge in his systems philosophy attempts to give a purely physiological and mathematically rigorous analysis of the concepts of health and disease, for both whole organisms and their constituent cells. Bunge[2:86] lists thirteen conditions which he sees as being necessary and sufficient conditions to distinguish living things from all other things. Logically this is flawed: either all living things are by definition healthy - thus satisfying his thirteen conditions in postulate 3.2 - or we are not speaking about a living thing. Whatever 'health' means, it is clear that a living thing may without contradiction be classified as 'unhealthy'.

A further approach developed within this perspective considers disease to be any entity or condition, the characteristic of it being that it deviates from the norm of the species, thus placing the organism at a biological disadvantage.[18] The most popular contemporary analyses of 'biological disadvantage' are socio-biological accounts of the lowering of inclusive fitness. Within this context, the approach is both too broad and too narrow. It is too narrow insofar as any 'disease', such as an infection which does not affect genetic fitness is not regarded as a disease. It is too broad insofar as voluntary celibacy and a life-style spent away from genetic relatives - and hence having no impact on their inclusive fitness - becomes a disease.

The second problem with such approaches is with the concept of normality. There are, as Vacha has pointed out, at least seven distinct senses of the term 'normality'.[19] (1) Commonness, usualness, in a statistical sense of lying within the range of variability of a double value of standard deviation on either side of a normal Gaussian curve; (2) averageness, that is, that which has the highest frequency of occurrence; (3) typicalness, that is conforming to some standard of typicality; (4) attaining adequate performance; (5) attaining optimal performance; and (7) a naive conception based upon a number of the preceding conceptions. Most medical accounts of 'normality' make use of either (1) or (2).

The difficulty with statistical concepts of normality as explications of disease is that they are difficult to operationalise. For example, consider whole populations affected by epidemics of the plague or parasitic infections. Using the statistical concept of normality, it is now logically impossible to classify the population as diseased. Similar difficulties are met in utilising an explanatory framework of normal function, that is concepts of disease which are based on the functions of parts failing to contribute to the goals for the whole organism. On this account of disease it is not possible for an organ to be non-diseased, and yet lose its function. As Margolis [20: 247] has pointed out, this is quite possible. For example, with increasing technological innovation in artificial reproduction, human sexual organs may completely lose their function of reproduction.

An awareness that concepts of disease are inextricably linked with judgements of a moral and ethical nature - that is, are a product of the social rather than the biological, is clear in the literature - especially when the subject of discussion is sexual behaviour[21], gender identity[22], race[23], addiction[24] or gender preference [25]. Authors such as Toon [26] seek to take into consideration the findings of this type of research, arguing that a distinction must be made between those disease categories which value judgements enter into and those which they do not. In other words, he wishes to distinguish problematic, social action which may be falsely classified as disease from real disease entities. Real disease patterns are those occurrences which can be analysed by the anatomico-physiological model. The argument is not sustained however. It cannot be argued that there are 'real' diseases uncovered by scientific medicine, and 'unreal' diseases which are ethical or moral judgements independent of medicine, since these latter, for their status depend on the former.

Indeed within contemporary medical philosophy, it has become almost impossible to distinguish disease from non-disease. Such an attempt to distinguish between disease and subjective psychological factors has been mounted by Kraupl-Taylor and Scadding.[6] They postulate that a disease in general - that which separates the class of patients from non-patients - is distinguished by: (1) a desire for therapy by the patient; (2) that others in the environment feel that aid should be administered; and (3) a concern expressed by a medical practitioner. However, as they acknowledge, these characteristics embody those social and cultural influences he wishes to rule out. They conclude:

> The dilemma is insoluble at present as there are as yet no hard and fast rules which are satisfactory enough to put the diagnosis of disease in general on an objective unassailable basis.[6:423]

This attempt to locate diseases as entities separate from social circumstances leaves theorists with inexplicable phenomena. Thus Bollet is taken aback by the fact that tuberculosis steadily declined after 1855 in the United Kingdom: that is to say, prior to the discovery of the tubercle baccillus.[27] By conceptualising disease as a biological entity the puzzling conclusion, to Bollet, is that 'many changes have occurred in the nature, frequency and distribution of major diseases, beyond those which can be attributed to improved medical understanding and use of diagnostic terms'.[27:15] Similarly Klepinger [28:581] can only conclude 'Some diseases change their expression; new diseases arise and some die out'. This inability to theorise disease only arises if diseases are given a privileged epistemological status, and not seen to be dynamically interrelated with social structures. The problem does not arise if disease action and categories are seen to be dynamic historical categories: more fully, if the self understanding of an epoch's illnesses, the action surrounding a disease and the social structure within which they exist are interrelated, as for example in Zinsser's classic study of typhoid, then these paradoxes do not arise.[29] Diseases in this perspective are social conditions.

We are now in a much better position to propose a sociological discussion of the question, what is disease? Perhaps the question should be framed more correctly: under what circumstances do some human actions become diseases - i.e. become human events, which are held to be closed off from human action? To establish the premise of this statement, that disease categories are the product of human action and not human (natural) events we can briefly review a book by Ludwick Fleck, first published in 1935.[30] Fleck's work is one of the most important early contributions to the sociology of medicine. In its explicit concern for epistemological issues it develops themes which have only very recently been revived in the discipline.[31,32] Starting from the premise of the sociology of knowledge, that 'a fact always occurs in the context of the history of thought and is always the result of a definite thought style' [30:95] Fleck examines the transition of the concept of syphilis from that of a carnal scourge, to a theological/ethical/mystical disease entity and finally to an empirical therapeutic disease entity. His central target in this discussion is bacteriology, that is, explanations of disease which have recourse to micro-organisms as the sole aetiological component. He points to three developments within medicine which undermine the concept of disease entities: (a) the discovery that some people carry germs without being sick; (b) the variability of micro-organisms; and (c) the theory of the filterable virus. He concludes that infection via invasion of the causative agent is the exception rather than the rule. He points to

> a divergence between the development of the concept of any disease and that of its causative micro-organism..... The presence of a micro-organism is therefore not identical with its host feeling ill. Consequently, the idea of a causative agent has lost the overriding importance it enjoyed during the classical period of bacteriology ... Today it can be claimed almost with impunity that the causative agent is but one symptom, and not even the most important, among several indicative of disease.[30:18]

How then to explain the transition between one explanation of disease to another - say from 'a primitive belief in demons through the idea of a disease miasma, to the theory of the pathogenic agent'?[30:100] Prefiguring Bachelard's[33] concept of epistemological breaks in cultural patterns of thought and Kuhn's[34] notion of paradigm, Fleck points to thought collectives citing Durkheim and Mauss' *Primitive Classification* [35] pointing out that as sociologists they 'exhibit excessive respect, bordering on pious reverence, for scientific facts'.[30:47] Fleck suggests that the maintenance of a particular concept is a social event, dependent not on individuals, but the individual membership of a group. If this is true as a general principle, he argues, then it is as equally applicable to concepts in our own society as it is in primitive societies.

The characteristics of medicine as a thought collective are twofold and contradictory. In his example of the case of syphilis he argues that it is seen as both carnal scourge, an ethical mythical disease entity and as an empirical therapeutic disease entity. On the first score, he argues, the devil 'haunts the scientific speciality to its very depths'[30:117] and on the second, a primitive image of war. The concept of infectious disease

> is based on the notion of the organism as a closed unit and of hostile causative agents invading it. The causative agent produces a bad effect (attack). The organism responds with a reaction (defence). This results in a conflict which is taken to be the essence of disease.[30:59]

The two factors come together in 'the disease demon' (cf. Sontag[36:69] on the military imagery of chemowarfare). Hence we could state Fleck's argument as follows: The production of concepts to explain human illness is an historically specific occurrence. There are then no 'facts' - but there are the genesis and development of scientific facts.

> Both thinking and facts are changeable, if only because changes in thinking manifest themselves in changed facts. Conversely, fundamentally new facts can be discovered through new thinking.[30:51]

Without developing any specific argument of his own, we can nevertheless, see the connection implied between military metaphors and nineteenth century imperial expansion, and the contribution of the heritage of Catholicism to our thinking. On the basis of his discussion Fleck concludes

> In science, just as in art and in life, only that which is true to culture is true to nature.[30:35]

To be clear on what is at issue here let us restate the premise leading up to our discussion of Fleck. The premise is that all human events are actions not behaviours; within our society this point is obscured by our distinction between nature and culture. It is sociology's task - in large part - to examine the development and maintenance of this distinction (a distinction which is most easily located historically in Descartes' mind-body dualism). The full implication of this position is that there are only historically specific classificatory schemes, which are the product of human interaction. Therefore, all explanations of human actions - including the cognitive facilities by which we explain our world - are open to sociological explanation. Hence science and medicine are inadequate explanations since they refer not to human actions but to behaviour and concepts of nature which are social constructs.

Sociologically then we can predict that the category of disease will come into play in those situations in which the boundary between nature and culture is problematic; and where agency and structure (or individual and society) are contested issues. In other words it will come into play in those situations where human events can only be problematically defined as either behaviour or actions. The corollary of this position is that when we find the concept of disease being operationalised we should look for these structures, forewarned that the concept of disease is covering them and obscuring them. In this we both follow and depart from Mary Douglas' anthropological work *Purity and Danger* [37]. We can follow her formulation that reflection on disease

> involves reflection on the relation of order to disorder, being to non-being, form to formlessness, life to death. Whenever ideas of dirt are highly structured their analysis discloses a play upon such profound themes. [37:5]

That is, reflection on disease is reflection on the status of human agency, and its relationship to nature, the body, and the social structure within which all are embedded. On the other hand we cannot follow her ahistorical and anti sociological claim 'that the difference between pollution behaviour in one part of the world and another is only a matter of detail'. [37:35] Rather if the most general sociological principle that structure and culture are interrelated is true, there should be marked differences in the conceptualisation of disease (and the implicit theory of human agency hidden within) over time and between social organisations: a simple reflection which anthropologists have indeed, through comparative fieldwork, made clear. [38:99]

For example, historians of the nineteenth century's attempt to control cholera at an international level who overlook this point are at a loss to explain the seeming intransigence of participants at the International Sanitary Conferences. Thus Howard-Jones cites the report of the first conference, held in Paris in 1851, which 'excluded not only political questions but also any discussion of scientific theory'. [39:163] This, to him, puzzling phenomenon does not present itself to sociologists. As Bilson points out, international attempts to control cholera failed 'because local needs made different theories about the disease more or less attractive in different countries'. [40:429]

Conclusion

This paper has identified three approaches in our understanding of the question 'what is a disease?' The first, the empiricist, is supported by a Cartesian epistemology. Diseases exist independently of human culture and are identified by the physio- anatomical methods of modern medicine. To the extent that medicine practises as a physical science then it will progressively come to understand more about, and by corollary, be in a better position to eradicate disease. However this position is open to serious criticism. Diseases do not present themselves as 'facts'. Rather they are constituted by theories which by definition go beyond 'facts'. Even if this relativistically informed epistemology is disallowed, the empiricist approach is open to attack on its own grounds. It operates on an unexamined meta-theoretical notion of normalcy. It was shown that his auxilliary premise leads the Cartesian empiricist to label as disease phenomena - such as celibacy - that which on their own terms are not diseases, as diseases.

Approaching the topic from a different angle, and implicitly responding to the problems in the Cartesian stance, the Hegelian normative orientation to the concept of disease acknowledges the value laden aspects of the concept. There are, however, logical difficulties with this limited attempt to take the concept of society seriously. In the first instance the normative approach still retains the key premise of Cartesian thought: that if disease could be stripped of social contamination then it would stand revealed in itself. The efforts by Hegelian medical philosophers to find and maintain this position has led to an acknowledgement that it is impossible to separate disease from these cultural categories. Further it leaves the Hegelian normative perspective on disease incapable of dealing with key phenomena in the history of western disease patterns: that diseases are produced, and subsequently remit, for reasons to do with the social organisation of society rather than medical knowledge. By accepting that diseases have an independent existence from the society in which they occur these researchers cannot account for the decline in mortality rates from socially constructed relationships such as tuberculosis. As Zinsser established in his study of typhoid, germs are a necessary but not sufficient cause of the disease. What is necessary are the social conditions to sustain the diseases.

It is only when disease is conceptualised as a manifestation of the social that these dilemmas for Cartesian and Hegelian systems can be resolved. In the Nietzschean position, represented by Foucault, not only diseases but the very categories of thought on which they depend - germs and bodies for example - have to be seen as culturally and historically specific. As O. Tempkin has argued, if a social history of medicine is to be developed then the contents of medical knowledge, as the basis of our self understanding for health, illness and disease must be seen as a specific historical product. [41: 86-88] In this way it is clearer how diseases are socially constituted - syphilis for example - or why there are disagreements over the nature of disease - for example, the case of international attempts to control cholera in the nineteenth century.

Notes

1. Boorse, C. 'On the Distinction between Disease and Illness' *Philosophy and Public Affairs*, 5, 49-68, 1975.

2. Bunge, M. *Treatise on Basic Philosophy, Volume 4; Ontology II: A World of Systems.* D. Reidel, Dordrecht, 1979.

3. Hesse, M. *Revolutions and Reconstructions in the Philosophy of Science* Harvester Press, Sussex, 1980.

4. Feyerabend, P. *Against Method* New Left Books, London, 1975.

5. Feyerabend, P. *Science in a Free Society* New Left Books, London, 1978.

6. Taylor, F. Kraupl and Scadding, J.G. 'The Concepts of Disease' *Psychological Medicine*, 10, 419-424, 1980.

7. Taylor, F. Kraupl 'Disease, Concept and the Logic of Classes' *Br. Jn. Med. Psychol.* 54, 277-286, 1981.

8. Kellman, S. 'Social Organisation and the Meaning of Health' *Jn. Med. and Philosophy*, 5, 133-144, 1980.

9. Rorty, R. 'Foucault and Epistemology' in D. Couzens-Hoy (ed.) *Foucault: A Critical Reader,* Basil Blackwell: London, 1986.

10. Harre. R. *The Philosophies of Science* Oxford University Press, Oxford, 1985.

11. Alford, J. 'Medicine in the Middle Ages: The Theory of a Profession' *Centenn. Rev.*, 23, 377-396, 1979.

12. Metz, K. 'Social Thought and Social Statistics in the Early Nineteenth Century' *Int. Rev. Soc. Hist.* 29, 2, 254-73, 1984.

13. Tesh, S. 'Political Ideology and Public Health in the Nineteenth Century' *Int. Jn. Health Services*, 12, 321-342, 1982.

14. Foucault, M. *The Birth of the Clinic* (Trans. A.M. Sheridan Smith) Tavistock, London, 1973 and 'Nietzsche, Genealogy and History' in *Language, Countermemory, Practise* D.F. Bouchard (ed.) (D.F. Bouchard and S. Simon trans.) Blackwell, Oxford, 1977.

15. Turner, B.S. 'The Practises of Rationality: Michel Foucault, Medical History and Sociological Theory' in Fardon, R. (ed.) *Power and Knowledge*, Edinburgh University Press, Edinburgh, 1986.

16. Elias, N. 'Sociology of Knowledge: New Perspectives, Part 1' *Sociology*, 5, 149-69; and Sociology of Knowledge: New Perspectives, Part II', *Sociology*, 5, 355-71, 1972.

17. Boorse, C. 'Health as a Theoretical Concept' *Philosophy of Science*, 44, 542-73, 1977.

18. Campbell, E, Scadding, J. and Roberts, R. 'The Concept of Disease' *Brit. Med. Jn*, 2, 757-62, 1979.

19. Vacha, J. 'Biology and the Problem of Normality', *Scientia*, 113, 823-46, 1978.

20. Margolis, J. 'The Concept of Disease' *The Journal of Medicine and Philosophy*, 1, 238-255, 1976.

21. Engelhardt, H. 'The Disease of Masturbation: Values of the Concept of Disease' *Bulletin of the History of Medicine*, 48, 234-48, 1974.

22. Rosenberg, S. and Rosenburg, S. 'The Female Animal: Medical and Biological Views of Women and her Role in Nineteenth Century America' *The Journal of American History*, 60, 332-56, 1973.

23. Cartwright, S. 'Report on the Diseases and Physical Peculiarities of the Negro Race', *The New Orleans Medical and Surgical Journal*, May, 691-715, 1851.

24. Fort, J. 'Should the Morphine Habit be Classed as a Disease?' *The Texas Courier of Medicine*, 12, 293-297, 1895.

25. Green, R. 'Homosexuality as Mental Illness' *The International Journal of Psychiatry*, 10, 77-98, 1972.

26. Toon, P.O. 'Defining Disease - Classification must be distinguished from Evaluation' *Journal of Medical Ethics*, 7, 197-201, 1981.

27. Bollet, A. 'The Rise and Fall of Diseases' *Am. Jn. Med.*, 70, 12-16, 1981.

28. Klepinger, L. 'The Evolution of Human Diseases: New Findings and Problems' *Jn. Bio. Soc. Sci.*, 12, 481-486, 1980.

29. Zinsser, H. *Rats, Lice and History*, Little Brown, Boston, 1935.

30. Fleck, L. *The Genesis and Development of a Scientific Fact*, University of Chicago Press, Chicago, 1935, 1979 ed.

31. Figlio, K. 'Medical Mythology' *Sociology of Health and Illness*, 2, 335-340, 1980.

32. Treacher, A and Wright, P. (eds.) *The Problem of Medical Knowledge: Examining the Social Construction of Medicine*, Edinburgh University Press, Edinburgh, 1982.

33. Bachelard, G. *The Philosophy of No* (G.C. Waterson trans.) Orion Press, New York, 1968.

34. Kuhn, T. *The Structure of Scientific Revolutions*, Chicago University Press, Chicago.

35. Durkheim, E and Mauss, M. *Primitive Classification*, University of Chicago Press, Chicago, 1963.

36. Sontag, S. *Illness as Metaphor* Penguin, London 1983.

37. Douglas, M. *Purity and Danger*, Routledge and Kegan Paul, London, 1966.

38. Fabrega, H. 'Social and Cultural Perspectives on Disease' *Journal of Medical Philosophy*, 5, 99-108, 1980.

39. Howard-Jones, N. 'The Scientific Background of the International Sanitary Conference 1851-1938, Part 2' *W.H.O. Chronicle*, 28, 159-171, 1974.

40. Bilson, G. 'The First Epidemic of Asiatic Cholera in Lower Canada, 1832' *Medical History*, 21, 411-433, 1977.

41. Tempkin, O. 'The Meaning of Medicine in Historical Perspective' in *The Double Face of Janus*, John Hopkins University Press, Baltimore, 41-94, 1977.

8 Sociobiology, ethics and human nature

Lucy Frith

Introduction

Sociobiology is generally regarded as a rather crude form of biological determinism and its status as a scientific theory has often been called into question. Further, it is fequently cited as a clear example of how scientific theories can be used to legitimize political doctrines and societal arrangements. However, this paper will concentrate on the way sociobiological analysis is used to shore up various philosophical positions within ethics and human nature debates. Modern ethics is a discipline where sociobiology's influence is still prevalent and its biological findings accepted. Sociobiology gains its strength from building on a common sense intuition that certain characteristics are embedded in our biology, and from this acceptable statement infers that every facet of humanity is biologically based and that this biology is best explained by a sociobiological analysis. Philosophers such as Mary Midgley, see sociobiology as a valid expression of the innate side of humanity and this creates major problems in her theory of human nature. In their ethical writings, the sociobiologists construct moral systems based on what they consider to be human nature in its natural form, without the distortions of environmental conditioning. It is primarily their interpretation of what is 'natural' and innate within humanity, as well as the elevated role that they attribute to it, that is unwarranted.

The critique of sociobiology will be directed at two levels; first, that it is illegitimate to base a theory of human motivation on the comparison between humans and animals. This will address the issue of sociobiology as bad science:

is sociobiology's legitimacy acceptable within a conventional scientific framework? Second, building on this critique, the application of sociobiology to philosophical debates on human nature and ethics will be examined. The debates over the nature/nurture argument have been a staple philosophical topic for centuries and this paper does not attempt to provide a definitive answer to this question. However, it aims to argue that sociobiology is not an adequate elucidation of the innate side of humanity, whatever its role in human causation.

Biological determinism and its recent articulation sociobiology, have attempted to legitimize an individualistic conception of human nature by arguing that such a nature is fixed by the genes. All human behaviour and action is explained as a result of the biochemical properties of the cells of each individual. Society, in turn, is deemed to be governed by the sum of all individuals and their genes, hence on these terms, social phenomena can ultimately be explained by recourse to the genetic make-up of the population, embodying a firm reductionist position.

Biological determinism is the articulation of a predominant common sense view in society: there are some basic human characteristics that cannot be changed or influenced because they are in some sense 'natural'. This was exemplified in the Thatcher-Reagan populist beliefs of the New-Right, overtly illustrated in a statement made by Patrick Jenkins in 1980, when he was Minister for Social Services: 'Quite frankly, I don't think mothers have the same right to work as fathers. If the Lord had intended us to have equal rights to go to work, he wouldn't have created men and women. These are biological facts, young children do depend on their mothers.'[1] This belief in innate bio-social characteristics have formed the basis of certain social policy measures that were implemented by the Thatcher-Reagan administrations. The guiding principles of these policies were a return to family values and a firm belief in the nineteenth century adage, charity begins at home. These principles were built on the premise that people will not want to help and care for those they do not know. In this way, a state providing benefits and medical support will be fighting a basic human tendency to simply look after one's own. Within the internal logic of this approach collective support structures will never flourish because they go against the grain of human nature, a nature that is not programmed to look after those who are not their own kin.

Biological determinism itself is a complex concept that incorporates an essentially static conception of human nature which states that, no matter where or when an individual is brought up, there are certain fundamental traits that they will exhibit and which cannot be changed. A biological determinist would argue that characteristics like aggression and territorial reaction to strangers, that are often seen as anti-social, are inevitable and such a move to jettison them in the interests of communal living would force people to behave unnaturally, and this could endanger the survival of the species.[2] This position has it roots in the conception of human nature expounded by theorists such as Hobbes, who saw society as a mechanism to regulate the inevitable features of human nature. These features were, in essence, selfishness, competitiveness and an overriding desire for glory and power. For Hobbes, man's dominant motivation was to attain power: 'All passions may be reduced to the Desire of Power.' Man did not want to co-operate, but simply strove to gain power. In this way the natural state was one of war. In Hobbes words: 'every man against every man.' Therefore men had to be forced, in the interests of survival, to co-operate with one another and live communally. Some followers of Hobbes viewed this state of affairs as inevitable because it was all encoded in biology and hence unchangeable.

Modern biology has built these intuitions into a scientific theory that aims to provide a greater understanding of humanity in all its aspects. As Barash states: 'Sociobiology is a whole new way of looking at behavior. It is the application of evolutionary biology to social behavior, an approach that has proven successful in animal studies and that may hold promise for a greater understanding of human behavior as well.'[3] It is conceived as a scientific theory and thus lends itself to repudiation or validation on scientific and empirical grounds.[4]

Sociobiology: an outline

The purpose of this section is not to give a comprehensive analysis of the contentions of sociobiology, but to develop the argument that it is a form of biological determinism and hence can be criticized on similar grounds to determinist theories. The major work in the field of sociobiology, published in 1975 to much popular acclaim, is E.O.Wilson's *Sociobiology: The New Synthesis*, in which he reiterates the basic premises of biological determinism. Namely, all aspects of human culture and behaviour have been coded in the genes, which in turn have been moulded by natural selection. For sociobiologists the driving force behind human motivation is the maximization of reproductive fitness and all behaviour is ultimately geared towards this goal. So an organism that has reproduced its genes and continues its existence beyond its own lifetime, has fulfilled its evolutionary destiny.

The conceptual basis of sociobiology is an adherence to the doctrine of evolution and natural selection and the reduction of complex entities to their component parts. These will be dealt with in turn.

Wilson states that his work, 'makes an attempt to codify sociobiology into a branch of evolutionary biology.'[5] He concentrates on the genotypic level of explanation rather than the phenotypic level. The phenotypic level analyses the whole of the organism, as opposed to simply analysing its genetic make-up, which is the province of the genotypic level. Sociobiology concentrates its analysis on the study of this genetic make-up, arguing that the genes that survive are the ones that are the most adaptive and the ones that ensure the maximum reproductive capacity of the species. To concentrate solely on this level can create problems, some of which Wilson recognizes. The most pertinent of these is the individual nature of the genotypic level; essentially natural selection is seen as an individual phenomena, one type of gene competing with another in order to preserve characteristics helpful to the organism. When viewed in this way, natural selection provides no room for any concept of altruism, a trait that both animals and humans demonstrably possess. That people help one another when all they should be concerned about is the survival of themselves is what Wilson sees as: 'The central theoretical problem of sociobiology.'[6]

The problem arises as a result of this formulation of the doctrine of natural selection, which incorporates the notion of individual organisms or genes competing and this predominates over any form of group selection mechanisms. Wilson tries to circumvent this problem by arguing that altruism is enlightened self-interest rather than any genuine feeling of care for other beings, this is not an original formulation by Wilson but stems from Darwin. In suggesting that by helping others one helps oneself, Wilson is able to keep the individual notion of gene selection while accounting for collective traits which might, at first sight, appear to contradict the individual genetic theory of natural selection.

This leads to the second conceptual basis of sociobiology, its reductionism. Reductionism stems from the ontological premise that parts are prior to wholes and therefore explanation must be concentrated at the base level of the individual parts, which in the case of organisms is the gene. As the authors of *Not in Our Genes* state:

> Sociobiology is a reductionist, biological determinist explanation of human existence. Its adherents claim, first, that the details of present and past social arrangements are the inevitable manifestations of the specific action of genes. Second, they argue that the particular genes that lie at the basis of human society have been selected in evolution because the traits they determine result in higher reproductive fitness of the individuals who carry them.[7]

Thus, all behaviour is determined by the various genes that make up an individual and an organism's guiding motivation will not be found unless the organism is reduced to its component parts.

To illustrate how sociobiology explains behaviour and motivation by a recourse to the genetic make-up of an individual, a sociobiological explanation of the reproductive behaviour of humans is an interesting area to consider. This is a crucial topic for sociobiologists as within their system it is the primary motivation of organisms to maximize fitness. Under sociobiology's formulation of reproductive strategies, men and women have radically different aims and goals within the context of maximizing their fitness. From the basic fact that the female maintains the fertilized egg, and the egg is bigger than the sperm, sociobiologists infer that she has the greatest investment in the process and a whole conceptual schema results. This is eloquently summed up by Dawkins:

> Females can be expected to invest more in children than males, not only at the outset, but throughout development. So, in mammals for example, it is the female who incubates the foetus in her own body, the female who makes the milk to suckle it when it is born, the female who bears the brunt of the load of bringing it up and protecting it. The female sex is exploited, and the fundamental evolutionary basis for the exploitation is the fact that eggs are larger than sperm.[8]

Sociobiologists assert that as the female has a greater investment in the child because (in mammals) she carries it for the gestation period, she will be more likely to look after it when it is born. The woman also has to implement certain procedures to make sure the man will assist in the child-rearing process. Either she chooses a fit, healthy, vigorous man to ensure that she has fit children who will be less dependent on the father, or she tries to entrap the man with the promise of domestic bliss to prevent him from abandoning her. Whichever option she chooses the woman is seen as a scheming, manipulative individual trying to get the man to do something that goes against his natural biological impulses. As Barash puts it: 'A man's sperm represents very little investment; a women's egg, while small itself, may have enormous significance to her...women should be more fussy than men.' So women should go for quality and men for quantity drawing the same conclusion as Dawkins: 'In effect I am suggesting a potential biological basis for the double standard.'[9]

This is exemplified in D.Symons'[10] account of human sexuality, where he discusses copulation as a female service to the male. Symons argues that it is always males that have to ask for and pursue sex and that men traditionally give presents to women whom they are attempting to attract, rather than vice versa. While women can always find willing sexual partners, men have to beg. Symons puts this down, not as a result of the conventional roles of women, but as a result of an inevitable biology. Hence: 'Copulation as a female service is easily explained in terms of ultimate causation: since the minimum male parental investment is almost zero, males stand to benefit from copulating with any fertile female (if the risk is low enough), where as females do not stand to benefit reproductively from copulating with many males no matter what the risk is.'[11]

According to Symons, men are also sexually aroused more easily and readily than women.[12] In order to bolster his claim, Symons cites pornography as an example of a situation where men are aroused by something women find repugnant: 'The most striking feature of pornotopia is that sex is sheer lust and physical gratification, devoid of more tender feelings and encumbering relationships, in which women are always aroused, or at least easily arousable, and ultimately are always willing. There is no evidence that a similar female fantasy world exists.'[13] Leaving aside for the time being this glaringly one-dimensional analysis of the nature of pornography, Symons argues that there is no female equivalent to this type of sexual stimulation and from this undisputed contention infers an illegitimate conclusion, namely, that the reason for this lack of female pornography is that women do not become sexually aroused by just looking at men, a situation which he claims is universal across all cultures.

There is no clear evidence to support Symons' claim that men are more easily aroused than women by visual stimulation. To take pornography as an example hardly makes a conclusive case, as it can be argued that pornography titillates men purely by its exploitative nature of women.[14] The supposed difference in arousal is justified by Symons on the grounds that it has a basis in the biological functionings of the sexes. The reasons for this discrepancy in visual sexual arousal between the sexes is due to the differing investment in the reproductive process, and so Symons states: 'A female's reproductive success would be seriously compromised by the propensity to be sexually aroused by the sight of males.'[15]

Symons' work shows how sociobiology aims to explain human behaviour by showing what is biologically possible. The existence of a certain gene, which determines certain characteristics, will if the characteristics are successful, ensure the evolutionary survival of the gene. For Wilson the fundamental causation lies with the human genotype and he predicts that: 'The genetic bias is intense enough to cause a substantial division of labour even in the most free and most egalitarian of future societies....Even with identical education and equal access to professions, men are likely to continue to play a disproportionate role in political life, business and science.'[16] In this way, according to the sociobiologists, measures to secure an equal society are fruitless due to the indisputable fact that we are not born equal and nothing can change that.

Sociobiology appears to be a clear example of biological determinism in that it reduces all behaviour to biological imperatives. Those writers who are generally more sympathetic to the claims of sociobiology,[17] have tended to point out that it does not ignore the influences of culture and stress the important position it is given in the work of the sociobiologists. Midgley states: 'All of them from Lorenz to Wilson, have stressed the immense importance of culture and have simply asked that both sets of causes should be attended to....And Wilson leaves us in no doubt that, when he thinks about it, he fully recognizes the importance to cultural causes.'[18] This argument encapsulates the view that expressions such as biological determinism have been wrongly applied to sociobiology; sociobiologists do not necessarily have to give up believing in social causes just because they are dealing with genetic ones as well.

In essence it may be true that a belief in genetic causes does not necessarily preclude cultural influences, but sociobiology is couched in terms that give rise to the primacy of genetic causes. This primacy is made inevitable by placing the genetic and innate causes at the beginning of the explanation process, a process that is necessitated by the reductionist methodology which sociobiologists use to explain phenomena. Society and culture must be considered but they are examined not as entities in their own right but as tangible products of the workings of genetic evolution. To defend sociobiology on the grounds that it gives equal weight to cultural factors is to misunderstand what sociobiology sets out to do, and to implicitly accept the one-dimensional analysis produced by simply considering genetic and biological processes. Sociobiology has to be considered on the grounds that the sociobiologists themselves state as their basic arguments, that the causation of human behaviour lies with the genetic make-up of the individual and that all societal wholes can be explained by reducing them to the genetic patterns that make them up.

Sociobiology: re-examined

The critique of sociobiology will be directed at two main areas: first, what will be referred to as the strong interpretation of sociobiology. That is, biological determinism formalized into a scientific theory, in which actions and traits are capable of reduction to our genetic make-up and such a genetic make-up capable of being analysed scientifically to provide an explanation of human nature. As sociobiology sets itself up as a scientific theory the critique will be directed against such claims. Second, what is called the weak interpretation of sociobiology will be criticized, this position recognizes that other factors such as culture, environment and society can influence individuals and play a part in their development while keeping the notion of the essential biological basis of human action as formulated by the sociobiologists. The weak interpretation is a far more subtle version than the biologically-based theory offered by Wilson, it incorporates a common sense approach to human nature, one of seeing it as influenced by culture but having basic core traits we cannot change. This weak form of sociobiology has been incorporated into philosophical debates on human nature most notably by M.Midgley and it is her account of this issue that is considered.

To take the strong interpretation first. As mentioned above, one of the major justifications given for the sociobiological position is its supposed scientific validity. This position will be judged by examining its methodology and its conceptual basis, to see if if can adequately lay claim to scientific rigour. Sociobiology sets itself up as a scientific theory and thus the critique in this section will be directed to establishing sociobiology as a form of bad science.

One of the major flaws in sociobiology is the ethnocentricity of behavioural description. This takes the form of giving metaphors the status of having a real identity, a mistake that is easily made in the field of animal-human comparisons. Animal behaviour is defined in the same terms as human traits or organizations and then the origin of these metaphors is forgotten and reapplied to humans. For instance, concepts such as slavery, aggression and tribalism are seen as actually existing entities having a concrete reality and a universality which can be applied to animal behaviour, in the belief that animals actually exhibit these traits. Once animals are seen to be exhibiting forms of social organisation that conform to these metaphors, they are inferred to be natural and reapplied to humans.

Sociobiologists claim that animal behaviour fits into these categories without realizing that they may have no existence outside the minds of white western scientists. Traits like entrepreneurialism are culturally specific terms that can hardly be said to exist cross culturally, never mind across different species. Barash, for example, talks about ants making slaves out of other species, but the concept of slavery on a wide scale implies some sort of political and socio-economic organisation that ants patently do not possess. Many of the supposedly universal human traits that sociobiologists constantly allude to are remarkedly similar to those displayed in the white western world. The investigators seem to be holding themselves up as reference points, not taking into account the cultural relativity of their position.

This makes it very easy for confusion to arise when making comparisons between animals and human beings. Once the arbitrary categories have been set up as so-called real objective entities, animal behaviour is fitted neatly into this explanatory framework. This makes sociobiology, according to R.Bleier, a self-fulfilling prophecy.[19] It is not possible to interpret animal behaviour in the light of what is believed about human behaviour and then use the interpretation of that animal behaviour to explain something about human behaviour. This is a circular argument. Barash gives a poignant example of this when he claims that plants commit rape. The male bombards the female plant with pollen as a means of reproduction and Barash infers from this that such a 'bombardment' is natural for humans too: 'Plants that commit rape...are following evolutionary strategies that maximize their fitness....We human beings like to think we are different...but we may have to open our minds and admit the possibility that our need to maximize our fitness may be whispering somewhere deep within us and that, know it or not, most of the time we are heeding these whisperings.'[20]

To equate the bombardment of pollen with rape in the human sense is quite outrageous. Rape has long been seen, not as the normal outlet of sexual frustration, but as a matter of power and insecurity. Although there is much doubt as to the causes of rape, it certainly cannot be seen as a natural 'whispering within' to use Barash's terminology. This implies some inevitability that presumably must be accommodated, since an eradication of rape would be a denial of natural humanity.

A further criticism of sociobiology is that in order to shore up certain positions, sociobiologists have been accused of ignoring unwelcome animal data; the Emperor penguins provide a good example of such findings. Their child-rearing strategy involves the male in a crucial role, as they incubate the eggs and the couple share the upbringing in an egalitarian fashion unheard of by sociobiologists. However, there is little point in entering into a war of counter-examples, as substantial anthropological research has adequately fulfilled this task.[21] The problem concerns the criteria sociobiologists employ to uphold their thesis. If their approach omits data unsympathetic to their cause and does not adhere to a rigorous scientific approach then their theory is untenable by their own criteria for validation.

The final criticism in this section relates to another error of description made by sociobiologists. This concerns their use of arbitrary agglomeration which strikes at the heart of their claim to scientific objectivity. Sociobiologists have no scientific basis for the areas they take as cogent units and arbitrary distinctions are made and dressed up as objective fact. There is a difficult problem in evolutionary theory of how to decide which parts of an organism one takes as the evolutionary unit. The authors of *Not in Our Genes* make the point: 'What is the correct topology of description, the natural lines along which the phenotype of the individual is to be divided for the purposes of evolutionary theory.'[22] There is no *a priori* way to decide which level or levels of description should be used to describe the evolutionary process. Should one speak of evolution of the hand for example, or is that too small a unit to be useful or is it too large? Should units such as the thumb be considered? When evaluating the evolutionary process it is important to take into consideration the pressures of natural selection on all parts of the body as to simply analyse one area is often not enough.

When it comes to analysing behaviour there are even more problems, as the same authors point out:

> If it is difficult to decide how to divide up the anatomy of an organism for evolutionary explanation, how much more care must be used for behavior, especially in a social organism?....Integrated cognitive function remains mysterious in its organisation, yet sociobiologists find no problem at all in dividing up all human culture into distinct evolving units.[23]

Whenever presented with fundamental conceptual problems sociobiologists manage to side-step the issue, seemingly unaware of what they have done. If there are problems in dividing up the human body into evolutionary units, then the problems of applying this analysis to an entity as complex as society would seem immense and ones which are not adequately dealt with by the sociobiologists.

Sociobiology appears unable to provide an adequate scientific basis for its theorizing. A justification for the validity of applying animal behaviour as an analogy for human actions is never provided, and the circularity this seems to imply is left unexplained. The formulation of categories of description without recourse to one's own culture is an almost impossible task and the difficulties in such an approach are never acknowledged by the sociobiologists. Their universalism, both for the behaviour of different cultures and species, is a tenuous position to uphold and one for which there is little validation. The sociobiologists do not seem to be able to give adequate responses to these criticisms. The essential justification for sociobiology, namely taking animal behaviour as the paradigm of uncontaminated behaviour, untouched by cultural influences, and applying it to humans, seems inescapably problematic.

Sociobiology and ethics

A further elucidation of sociobiology's dubious scientific status can be seen in its treatment of the moral issues that arise in the application of biological findings to society. At first it would appear that sociobiologists should have nothing to say about ethical obligations and imperatives, their function is to simply report on what they see, using appropriately rigorous scientific methodology. By the traditional scientific nature of their work it could be argued that a discipline such as ethics is beyond their range since it concerns qualitative rather than quantitative aspects of human existence. But quite the converse has happened writers, such as Wilson, have devoted much time and energy to the project of incorporating ethical premises into their biological theories,[24] and to proving how their biological findings can influence the grounding of ethics, arguing for the reappraisal of traditional mores in the light of these new developments.

An analysis of the treatment of ethics by sociobiologists clearly draws out the assumptions about human nature that underlie their theories and the purposes that such a biological determinist understanding of human nature are formulated to ensure. Wilson's writings on ethics provide an illustration of these trends in sociobiology. A brief outline of his ethical system will be given and P.Singer's account of the relevance to ethics of a biological determinist understanding of human nature in *The Expanding Circle* is considered, particularly his illuminating critique of Wilson's position.

Wilson's ethical system begins by gliding over conventional barriers between fact and value. He posits sociobiology as a theory that sets out to provide answers to questions of a normative nature, such as how society should be organized, as well as questions attempting to operate in a descriptive fashion, showing how the world is. In this way Wilson is able to argue that sociobiology could provide the scientific rationale behind new and better ways of organizing societal arrangements. This discipline could, from factual description, infer the rights and wrongs of how society should be organized. Hence:

> We do not know how many of the most valued qualities are linked genetically to the more obsolete, destructive ones. Co-operativeness toward

groupmates might be coupled with aggressivity toward strangers, creativeness with a desire to own and dominate, athletic zeal with a tendency to violent response and so on.... If the planned society - the creation of which seems inevitable in the coming century - were to deliberately steer its members past those stresses and conflicts that once gave the destructive phenotypes their Darwinian edge, the other phenotypes might dwindle with them. In this, the ultimate genetic sense, social control would rob man of his humanity.[25]

Wilson argues that future societies could be organized by rigorous planning and this would be workable because knowledge of the genetic make-up would make people amenable to manipulation. With the advanced techniques of sociobiology, it would be feasible first, to find out what exactly constituted the inherent biological base of people's ethical beliefs and second, to organize society around these findings. Wilson contends that all human traits, ranging from those individually manifested such as aggression, possession over one's own property and emotions, to those carried out at a societal level such as genocide, war, religion and taxation, for example, can be adequately explained by biology. Socially manifested traits can be explained by reducing them to traits exhibited in individuals and in turn reducing them to the products of an individual's genetic make-up. Once this task of charting the genetic make-up of humans is extensively completed, then biology can show us the pathways that humanity should take in the future.

This prescriptive side of sociobiology is most apparent when Wilson is considering the realm of ethics. It is often hard to tell when he is describing a given situation or being normative, as the two are often not clearly distinguished. He begins his ethical discussion by claiming that there is a clear common core of ethical standards and judgments throughout human societies. From this he infers that there is a biological basis to this that would account for continuity between cultures which must have been inherited from our prehuman ancestors. Thus, our ethical judgments are not culturally specific creations but the outcome of an inevitable biology. Ethical judgments become analogous to other biological imperatives like the need to eat and sleep; we can no more choose them than we can choose to never nourish ourselves again.

Sociobiology enters the realm of ethics when it considers the development of altruism and this concept provides a mechanism by which human behaviour can be applied to animals. It would be impossible to judge whether a lion was behaving ethically or not, but in the domain of sociobiology it is possible to document altruistic acts by non-humans. These altruistic acts provide the basis for human ethical decisions, the altruistic act of protecting one's own family becomes entrenched in ethical codes. The proposition that there is an ethical duty to support your own family above others and look after your own is compared to altruism in animals and given a biological basis. Hence, ethical codes, via altruism, are reducible to innate biological factors. As Singer states: 'Understanding the development of altruism in animals will improve our understanding of the development of ethics in human beings, for our present ethical systems have their roots in the altruistic behaviour of our early human and prehuman ancestors.'[26]

Singer's evaluation of Wilson's position and his analysis of the merits of a biological basis for ethics provides an illuminating account of the issues involved in formulating such a grounding. Singer outlines Wilson's theorizing on ethics, which is an attempt to transform it into a more rational discipline, founded on fact rather than value. A transformation which can be adequately affected by a sociobiological explanation of human nature. Ethics would be based on inherent biological facts about human nature: what is good for people would be amenable to scientific analysis, once ethics had been removed from the area of speculative judgment into one of concrete certainty.

In order to evaluate the feasibility of grounding ethical codes in the inherent biology of a subject, Singer divides up Wilson's application of scientific findings to ethics into three categories: first, science could produce new knowledge that would enable us to be more informed about the consequences of our actions; second, biology could transform ethics by undermining existing ethical beliefs; and finally, science could provide us with the new ethical premises themselves. By distinguishing the various ways in which science can be applied to ethical judgments, Singer attempts to separate Wilson's viable findings from the unjustifiable ones. Singer does not reject out of hand an attempt to provide a more scientific basis for ethics, just certain applications made by Wilson. These categories will be dealt with in turn.

First, Wilson argues that science could produce new knowledge that would enable us to be more informed about the consequences of our actions. This would prevent the situation arising where an ethical theory could result in consequences that the theorist would neither want nor predict. The added information would have the effect of simply clarifying the results that a certain action might entail. Once a detailed theory of human nature had been formulated with the aid of sociobiology, then the results of certain ethical judgments could be ascertained and whether these judgments benefitted people could be evaluated. The ethical judgment that it is wrong to kill one's children could be validated on the grounds that if all parents killed their children, then not only would the species decline, but the genes of the individual parents would not survive. This would not fit in with what sociobiologists define as the fundamental motivation of humanity, namely to maximize reproductive fitness. In this way, the moral maxim, it is wrong to kill one's children can be reduced to basic biological genotypes.

In Singer's eyes, this is a legitimate application of science to ethics, as it does not blur the distinction between fact and value. Singer states: 'Traditionally facts have been regarded as the domain of science, values as the domain of ethics. What consequences our actions will have is a question of fact....For sociobiology to tell us something about the ultimate consequences of our actions would not threaten this traditional division of territory between science and ethics.'[27] Further, this application would not affect the consequentialist position, which regards the consequence of the action as the important area for evaluation, not the action itself, and for that reason the core of the theory would remain unchanged. More information would simply alter the process by which the actions are chosen, as there would be more factors to consider when making an evaluation. Singer states:

> Since information about the consequences of our actions does not tell us which consequences to value ...most ethical theories simply incorporate new information about the consequence of our action into our ethical decisions in a way which does not effect the fundamental theory of value itself.[28]

In this way: 'The central question of ethics, the nature and justification of fundamental ethical values would remain untouched.'[29] Science would only affect the reasons for formulating a certain belief, not the belief itself, and would pose no threat to traditional distinctions between ethics and science.

This contention is obviously not what Wilson has in mind when he formulates his relationship between science and ethics, a point Singer does not recognize. Wilson sets out explicitly to blur this 'traditional division' between fact and value. Once there are scientifically proven objective facts about human nature, then the domain of ethics is no longer one of speculative value because we can know what is good for people and therefore what is right. Ethical decisions can be directly deduced from the scientific evidence of what is good for humans, for instance, it could be scientifically proved that Kant's ethic of inflexible moral rules will lead to some genetic or ecological disaster and this should invalidate the theory. This takes the brunt of the refutation of moral codes away from the subjective into a matter of verifiable fact.

The second application of science to ethics is the way in which biology could undermine existing ethical beliefs. This works on two levels, superficially any ethical belief that is based on an assumption of what is natural for human beings could be tested. Wilson would argue that this can be done by sociobiological techniques and it would be possible to tell with certainty what is natural and presumably what is natural is both good and desirable. As Singer states: 'Strictly speaking, the impact of biology here is not to render the ethical belief untenable, but to destroy the original justification for the belief.'[30] Therefore the basis for any ethical position can be justified or debunked on the grounds of better supporting evidence. This could be used to invalidate theories based on an appeal to natural law, but as this is no longer a widely held belief (outside religious circles) it does not have dramatic implications for ethical theory.

The more fundamental way in which biology could undermine ethical values is by fitting our belief structure into the context of our evolutionary development. In this way ethical values become the result of adaptations and part of our evolutionary history rather than moral absolutes. That is, all moral judgments are the product of evolution and could have taken vastly different forms than they do today. There are no longer high moral absolutes, all our moralizing has a biological and cultural specificity, and a more detailed knowledge of our biology can reveal this. This gives us a unique tool to criticize our moral institutions. As Singer states: 'Precisely because science is outside ethics, the scientific study of the origin of our ethical judgement is a fulcrum on which we can rest our critical lever. In itself, science cannot compel us to abandon a principle - a fulcrum is not a force - but coupled with a commitment to rationality, it can provide a leverage against basic ethical principles.'[31] Under this approach moral institutions can be evaluated and some dismissed as relics of our evolutionary history.

The exact way in which Wilson conceives the functioning of this approach is summed up by Singer:

> biological explanations of ethics discredit only those ethical principles they show to be relics of an earlier stage in our history, better suited perhaps to a tribal society than a modern urban life. Other ethical principles will be shown to be biological adaptations which remain well suited to the contemporary human situation. Those principles are justified by evolutionary theory. They will be the principles we retain.[32]

In the final category Wilson looks forward to the time when biology will make it possible for a better moral code to be selected, and such a moral code could be provided by science. Sociobiology, as Singer states, attempts to, 'discover ethical premises inherent in man's biological nature.'[28] Wilson argues that sociobiology will find traits and so forth that are already in man's nature but have just not been explicitly made known. This process brings up the crucial issue in Singer's discussion of Wilson's ethical theories - his treatment of the fact-value distinction. Singer argues that Wilson attempts to slide over this divide by formulating a theory on the grounds of what is and arguing that this is how it should be. This is an illegitimate move, in Singer's view, as the gap between facts and values is unbridgeable. Both Wilson's transcendence of this division and Singer's criticisms raise important problems for sociobiology.

Singer sums up the logical steps that sociobiologists take to deduce ethical premises from biological statements:

> As the sociobiologists say, we are evolved biological organisms and our brain and our emotions reflect the evolutionary adaptations that have

enabled us to survive. Our values and ethical systems are the product of our evolved nature. Isn't it then possible that as our knowledge of biology and physiology advance, they should come to reveal ethical premises inherent in our biological nature, thus bridge the gap between facts and values.[34]

As our brains and emotions reflect our evolutionary adaptations, and our ethical systems are the products of our evolved natures, then ethical premises could be found to be inherent in our biological natures.

Singer criticizes Wilson here for trying (illegitimately) to bridge the gap between fact and value. Singer argues that we choose our ethical premises, admittedly on the basis of facts, but not because of them. Sociobiologists in his view are confused as to the exact nature of ethics, seeing it as a discipline geared to making predictions or explaining action, whereas in fact it is one of puzzling over what we ought to do. Facts themselves have no direction so ethics directs, Singer says: 'In themselves facts have no direction they are neutral about what we ought to do.'[35] Ethics tells us what to do, whilst science provides the brute facts.

Singer argues: 'No science is ever going to discover ethical premises inherent in our biological nature, because ethical premises are not the kind of thing discovered by scientific investigation.'[36] In this way:

> Neither evolutionary theory, nor biology, nor science as a whole, can provide the ultimate premises of ethics. Biological explanations of ethics can only perform the negative role of making us think again about moral intuitions which we take to be self-evident moral truths but can be explained in evolutionary terms.[37]

But this is essentially Wilson's aim (and one that Singer does not seem to fully recognize) in linking ethics to sociobiology he wishes to overcome ethical subjectivism and give ethics a firm scientific basis. So mistakes as to what is good or right for people will no longer be made and a concrete ethical system can be formed in accordance with human nature.

This application of science to ethics illustrates the normative conclusions sociobiology has been formulated to reach. In many defences of sociobiology the claim has been made that it does not seek to prescribe any sort of behaviour or organisation of societies, it is merely painting a picture of the way the world is. This is how it is, rather than this is how it ought to be, or should be; but there is a very fine line between the two, one that Wilson, for instance, makes no effort to keep distinct. It is his treatment of the fact-value distinction that causes problems for his theories. One of the major validations for sociobiology is that it is a purely descriptive theory, and one that conducts its methodology along rigorous scientific lines. If this results in unpleasant conclusions, then it is a risk the theorists have to take and one they are not responsible for. Inherent in this approach is the fact-value distinction. Science deals specifically with the facts, any value judgments to be drawn from this bare data are the province of another discipline and not the concern of the scientist.

The objective scientific emphasis of sociobiology is one of the major planks in its validation. As a theory it is put in the context of wider proven hypotheses, such as evolutionary theory and sociobiology of the animal kingdom.[38] Sociobiology's case rests on the premise that it deals solely with facts and that these facts can be validated in an objective fashion. Within this traditional scientific context a theory is fundamentally flawed if it traverses the fact-value barrier. The long and complex debate over the validity of the fact-value distinction, will not be considered in any length.[39] The crucial point for this discussion is to put the problem of the fact-value distinction in the context of traditional scientific discourse, in which Wilson *et al* operate. Within this scientific framework, as it was argued above, it is illegitimate to slide over the distinction as Wilson does. Additionally, the dangers in dressing a value based theory up in the mantle of factual objectivity and resting the validation for it on objectivity, are obvious. Wilson never states that he infers values from his findings and his conclusions are always couched in terms of undisputable truths. If one is arguing from within a traditional scientific framework, it is this tacit bridging of the fact-value distinction, not admitting one might be making value judgments, that is an indefensible position. As the authors of *Not in Our Genes* commented:

> In *Sociobiology*, Wilson committed the naturalistic fallacy of 'the genetically accurate and hence completely fair code of ethics,' but shortly after, in *Human Decency is Animal* he cautioned against deriving 'ought' from 'is'. The effective political truth, however, is that 'is' abolishes 'ought'. To the extent that we are free to make ethical decisions that can be translated into practice, biology is irrelevant; to the extent that we are bound by our biology, ethical judgments are irrelevant.[40]

Wilson clearly illustrates this directional quality of sociobiology when he proceeds to argue about finding a new basis for ethics, a traditionally value-impregnated area of inquiry. Scientific technocrats will be able to organize future societies, making it clear that his vision of the future is one in which, with an extensive knowledge of human biology, scientists will be able to structure societies along certain indisputable lines. This will take some time, as biological knowledge is not sufficiently advanced to provide enough information about the human genotype to make such a project possible. Wilson states: 'A genetically accurate and hence completely fair code of ethics must wait.'[41] This prospect is very far away and the probability of it happening highly unlikely, but it does show that a defence of sociobiology as simply a descriptive theory is untenable.

Midgley and the weak interpretation of biological determinism

It is in its use by other disciplines that sociobiology has had its most persuasive and covert influences. Although it has been widely rejected by the biological community it has been retained within certain philosophical accounts. Mary Midgley attempts to explain human nature by a recourse to animal-human comparisons and provide a synthesis between biological causes of behaviour and environmentally conditioned aspects. Here sociobiology is unquestionably accepted as the correct explanation of the innate-biological facets of humanity, and it is this acceptance that leads her theory into problematic waters.

In *Beast and Man*, Midgley sets out to clarify our thinking on human nature, a task she argues that can only be adequately performed by philosophical inquiry. She aims to clear up various misconceptions that have arisen in the field of animal-human comparisons. In order to understand this reasoning, her theorizing on human nature must be put into the context of her wider philosophical system. Of particular importance is the role she gives to philosophical inquiry, namely one of erecting the conceptual framework for future discussion, 'building the conceptual framework for any study is a philosophical task, and that framework, once it is well built, can often be taken for granted in the later factual enquiries.' In this way, philosophy has a 'method-sorting function.'[42] For Midgley, philosophy is the erecting of conceptual skeletons upon which the flesh of facts and the clothing of values can be hung. This formulation of philosophy does not explicitly state that it is impartial and neutral, in fact, Midgley devotes much time and space to arguing the converse.[43] But this separation of philosophy from the specifics of enquiry can only lead to the discipline being regarded as different and independent of facts and values.[44] This notion of philosophy informs her conception of science and methods of general inquiry, namely that both can be objective. The problems this raises within her analysis of human nature will be considered later.

Midgley's formulation of human nature, as will be argued, is itself a branch of biological determinism. For clarity this will be called the weak interpretation of biological determinism. It is a position which encapsulates the basic premise of sociobiology, that our behaviour is inextricably linked to our biology but seeks to combine this with the notion of culture playing an equal but secondary role in causation. The first premise is upheld by the comparison of species argument used by the sociobiologists and the second is taken as a self-evident truth and used to fill in the gaps a purely biological understanding of human nature would leave. It is the weak interpretation that most clearly brings out the persuasiveness of the inherent concepts of human nature on which biological determinism builds. Being the less extreme form it is more easy to accept, although it must be realized that it rests on the same dangerous foundations as the strong thesis.

Midgley sums up her position in the following way:

> 1. Innate tendencies are extremely important in the formation of human behavior.
> 2. Recognition of these innate tendencies is not in the least inimical to human freedom. On the contrary, if there were no such tendencies, the concept of freedom would be unintelligible.
> 3. Careful and informed comparisons with the behavior of other species can illuminate human behavior. They enable it to be placed in its evolutionary context, and are an entirely proper and useful contribution to our understanding of it.[45]

Midgley's major concern is to stress the importance of biological factors in determining human attributes, while locating the reason for neglecting these biological factors as a desire to protect human free will. When evaluating the criticisms of Wilson's *Sociobiology: The new synthesis*, she states: 'the book has run into opposition of a political kind from people who believe that any notion of inborn active and social tendencies, if extended to man, threatens human freedom....The notion that we "have a nature", far from threatening the concept of freedom, is absolutely essential to it.'[46]

For this reason Midgley is strongly critical of theories of human nature that seek to minimize or ignore the effects of innate biological factors on human behaviour and characterizes them all under the heading of 'blank paper theories'. She traces the development of these theories beginning with Locke and culminating in J.B.Watson, the founder of behaviourism:

> Man, he declared (Watson) had no instincts. This mysterious news was remarkably well publicized; there seems to be nobody who studied any sort of social science in English speaking countries between the wars who was not taught it as gospel. Its obscurity, however, has made it increasingly a nuisance and no sort of help to inquiry. Not only do people evidently and constantly act and feel in ways to which they have never been conditioned, but the very idea that anything so complex as a human being could be totally plastic and structureless is unintelligible. Even if - which is absurd - people had no tendencies but the general ones to be docile, imitative and mercenary, those would still have to be innate, and there would have to be a substructure governing the relations among them.[47]

In this way the debate is crystallized into two polarities, one can either choose between being a 'blank paperist' or a biological determinist. Although she specifically formulates the discussion in this way, Midgley is strongly critical of this polarization, which she argues has become the dominate terminology in which the debate is couched. She puts forward an alternative view, arguing that what is needed is a synthesis of the two opposing explanations of human behaviour. Neither position alone can encompass all explanations of human behaviour as both leave out important elements which determine our actions. Hence: 'Everything that people do has its internal as well as its environmental aspect, and therefore its causes in the nature of man as well as outside him.' and, 'There is simply no need to take sides between the innate and outer factors in this way. We can study both.'[48]

So far so good. It seems a reasonable claim to make that a synthesis between two extreme versions of human nature is the best way forward in providing an adequate explanation of behaviour, but the mechanisms by which Midgley justifies this claim are invalid. Midgley's methodology leads to unwarranted consequences with detrimental implications for adequately theorizing the social construction of humans. The grounds upon which she bases her evaluation of human biology is at fault. The areas that need particular attention is her constant recourse to the work of various ethologists,[49] to substantiate the biological premises, and her adherence to the view that the animal comparison studies carried out by such ethologists are valid and can unproblematically be applied to humans.

She begins by arguing that we need to look outside the human race to understand human's motives and causation: 'Because, our cultures limit so subtly the questions that we can ask and reinforce so strongly our natural gift for self-deception.'[50] What is needed is to go outside human culture, outside the human species and look at animal behaviour. This can show the cultural specificity of actions, certain patterns and similarities that humans share with animals. Once animals have been studied, various common traits become apparent, which are often possible to see manifested in humanity. This basic method of human-animal comparison, according to Midgley, is an unproblematic methodology and generally accepted. She goes on to argue that these comparisons must be more rigorous: 'The further things which are needed, and which are now being vigorously developed, are a careful, thorough, disciplined procedure for making the original observations of animals precise and a subtler technique for comparison for checking the different sorts of variation in different species and linking them to their different sorts of causes.'[51] Accordingly, it will become possible to understand human behaviour by relating it to animals.

Midgley contends that essentially there is very little difference between the two species and this is her major argument for the claim that it is legitimate and profitable to compare human behaviour to that of animals. To prove this she argues that our conventional view of animals as irrational 'beasts' is an unjustified one: 'The popular notion of lawless cruelty which underlies such terms as brutal, bestial, beastly, animal desires and so on.' is used 'uncriticized as a contrast to illuminate the nature of man.'[52] Animals in fact can and do lead just as ordered and structured lives as humans. As an example she cites Lorenz and company's study of wolves, which found them to be paragons of steadiness and good conduct: "They pair for life, they are faithful and affectionate spouses and parents, they show great loyalty to their pack and great courage and persistence in the face of difficulties, they carefully respect one another's territories, keep their dens clean, and extremely seldom kill anything that they do not need for dinner.'[53] This study of wolves proves, according to Midgley, that to see animals as bestial is just transferring our worst traits on to them. This condemnation of animals was solely performed in order to uphold man as the highest form of species rather than having any factual basis.

The view that animals are terrible creatures, capable of great cruelty and incapable of mutual social existence can be traced back to Plato who saw man as having a beast within which was evil and irrational. Man throughout the ages has set himself up as morally superior, an island of order in a sea of chaos. Quite clearly such a formulation is unjustified, Midgley is right to criticize these perceptions of animals as the evil, crueler side of life forms. Many of the terrible traits humans attributed to animals were in fact solely human ones - no species of animal has practised mass genocide. It is a reasonable argument that humans have projected their own fears and desires on to animals.

Midgley then argues that it is only invalid to project certain traits on to animals. Taking her example of the wolves, it is acceptable to attribute them with orderliness, faithfulness and cleanliness, but not with cruelty nor irrationality. Lorenz's wolves seem to be paragons of humanity, exhibiting much more order, love and courage than most people ever do. This idea of a wolf's cosily organized life is subject to the same counter-arguments as the claim that humans project their worst fears on to animals. By stressing the sociability of wolves it would seem to follow that they may be subject to the same distorting factor as humans are, namely social conditioning. This could make them unreliable case studies under Midgley's schema.

In Midgley's analysis of animal behaviour little weight is given to the problems of anthropomorphism and ethnocentricity. She does, however, recognize the circular nature of such comparisons: 'The point of my discussion is to show how and in what cases comparisons between man and other species makes sense, but I must sometimes use such comparisons in the process. I think the circle will prove virtuous, however, if it abides by the following rule; comparisons make sense only when they are put in the context of the entire character of the species concerned and of the known principles governing resemblances between species.'[54] She states it is invalid to compare infanticide in hamsters with infanticide in humans, but gives no logical reason why this case differs from comparing territorial aggression in apes and humans, a comparison she regards as a fair. Following Midgley's logic, it is possible to argue that hamsters eat their young when, for what-ever reason, they are incapable of looking after them, and that humans too, are basically driven by this motive when parents who cannot cope nor look after their child batter it to death.

Midgley defends her position by claiming that the ethologists cannot be accused of falling into the trap of anthropomorphism. She does this by arguing that people are in just as much danger of making mistakes about other people's feelings as about animals. It is common to pronounce on others' feelings even if the views are conjectural: 'As Ryle pointed out in *The Concept of Mind*, none of us need suspect that (in spite of our constant success) we are constitutionally unable to pronounce on other people's feelings, that we are locked away in Cartesian solitude...most of us would be willing to accept that we can know something about human feelings.'[55] Midgley's argument is that the only problems humans have with pronouncing on animal feelings are the ones they encounter with their own species. She submits that it does not prevent them discussing human feelings, so why should it prevent them from considering the feelings of animals. This is not a justified counter-argument. The claim that it is possible to empathize with fellow humans is very different from claiming that we can impose our categories of explanation on to animal traits. There have been many other explanations of animal-human continuity which attempt to formulate this link in a very different way from the sociobiologists and have bearing on the empathization between species. Barbara Noske in a recent book,[56] has put forward an alternative thesis about the relationship between humans and animals that shows how a different approach would tackle such issues.

Noske's thesis is markedly different from many other theorists who have explored similar issues, she is prepared to state without reservation that there are great continuities between species. There are many traits that are traditionally thought to be solely facets of humanity (the higher order cognitive behavioural patterns for instance) and she argues that these are exhibited by various species. Animals in Noske's view are very complex sentient beings and the idea that they are merely machines or incapable of communication is an unjustifiable one, postulated to preserve the uniqueness of humanity. In her opinion animals fulfill many of the criteria of personhood; language, sociality, and an almost Hegelian sense of self in relation to others to name but a few. However, to argue that animals possess human traits is not the end of the argument for Noske; she contends that these continuities should only be used to provoke us into realizing that animals are really much more sentient than previously thought. In order to free animals from the object status, she argues, they have been given in western culture, their claim for autonomy cannot rest solely on arguing that they are the possessors of human attributes, their own animal subjectivity must be recognised and established. It is on this basis that their rights must be defended and the crucial issue for Noske. There is a dilemma in that there seems to be no alternative frameworks to impose on animals other than object status or human subject-status. What Noske wants to see is a notion of a non-human subject, that would provide an adequate evaluation of animal traits, while breaking free of the limitations of always defending animals on the grounds of their similarity to humans.

Noske's major contribution is in stepping outside the terms used in the debates over animal-human comparisons by sociobiologists and ethologists. Noske attempts to formulate a notion of animals subjectivity independently and without recourse to such comparisons. By trying to see things from the animals point of view (as far as she is able) a whole different perspective is created. It is in this respect that she differs radically from sociobiologists: she aims to create a specific animal subjectivity that will accept animals as equal but different, arguing that we cannot explain animal behaviour by a constant recourse to human actions.

In Noske's estimation it is not possible to see animal behaviour as a mirror of our own, they communicate and exist in communities, but these are not analogous to human society. Animals have their own species specific way of doing things, which, although it is much more complex than previously thought, it is not a diminutive version of human organiaztion. This analysis highlights problems in Midgley's ability to make such comparisons and provides a much more feasible view of animal behaviour.

Midgley's ultimate justification for animal-human comparisons is a reliance on scientific methodology to uncover the facts of the matter. 'In sum, whether and how far, interspecies communication works for feelings and motives is an empirical question.'[57] It is an empirical question that such people as Lorenz *et al* are able and qualified to answer.

The definition of the problem in empirical terms brings us to a further criticism of Midgley's position; she consistently takes issue with certain positions and then proceeds to reinstate them in her reasoning, simply inferring from them a different conclusion. An illustration of this can be seen when Midgley argues that it is not possible to project our traits on to animals in order to justify our moral superiority and then in the same breath argues that wolves exhibit numerous human traits in order to justify her thesis of human-animal compatibility.

A more damaging example of this reasoning is its application to science. To uphold her argument on animal-human comparisons, the ethologist becomes an objective purveyor of value-free, neutral, scientific method. When comparing the ethologist with Marx and Freud, she states: 'The ethologist, on the other hand, does not want to say that human nature is basically anything; he wants to see what it consists of....He proceeds more like a surveyor mapping a valley...he finds some of them tend to run together....If he finds an apparently isolated activity, with no connection with the creature's other habits, he simply accumulates information until a connection appears.'[58]

Midgley does not take into account any of the problems associated with this notion of a value-free, impartial ethologist, gathering observations and data, as spontaneously and as freely as one might gather flowers. She puts forward no conclusive arguments for the ethologist having any greater access to the true nature of animals and unquestionably accepts their findings as objective expositions of the truth. In Midgley's terms the ethologist is able to avoid the problems which non- biological accounts of human nature have encountered, that is, not having a concrete basis. Midgley never gives conclusive proof of the argument that ethologists have managed to circumvent problems encounted by other theories of human nature and puts forward no substantiation for the claim that ethology is an objective scientific discipline.

Midgley's fundamental position is that in order to understand human behaviour and motivation, a double-headed approach of both innate and environmental factors is valid. The problem lies in what she takes as the correct explanation of the innate biological side of causation. The fact that we have certain biological constraints is undeniable and bound to affect our behaviour, but to take the ethologist's version of a strong scientific explanation of our biology is unjustified and can lead to some bizarre notions of human nature. This is where her position comes closest to the sociobiologists, Wilson, for example does not deny the influence of cultural factors on human behaviour. The problems lie in the sociobiologist's definition of the innate and biological constituents of humanity. It would be better for Midgley not to rest her arguments on the interpretation of animal behaviour offered by the ethologists, as a reliance on this position weakens her arguments considerably.

To reinforce her position she often performs a sleight of hand when dealing with the relative merits of biological determinism and the rival claims of environmentalists. She emphasizes that Lorenz and Wilson *et al* take into consideration environmental factors as well as the biological. She gives the sociobiologist's position as one of synthesis between biology and the environment. She admits elements of synthesis into the ethologists position but is not so charitable with the environmentalists, always defining their position as the 'blank paper theory', an extreme form of environmentalism that does not admit that genetic factors have any influence on behaviour.[59] She thus weakens their case considerably by defining it in such a way and upholds the ethologists' position as the most sensible and coherent approach.

Midgley concludes that the most likely reason that theorists are so reluctant to admit genetic factors influence human behaviour is the fear that this would limit human freedom and imply some form of determinism. Now this is quite true as it would be terrible if the evils in our societies were unchangeable and predetermined. When people kill others it would be enough to say that they could not help their behaviour because they had been naturally programmed in this way. In order to avoid this conclusion, Midgley relies on her distinction between causality and fatalism arguing that: 'Causes do not force effects to happen - that is a superstitious view - but they do help us to understand and predict them. Perhaps then, if we are to understand it, we must for theoretical purposes think of human behaviour as predictable.'[60]

In defence of her position that determinism would not result from admitting genetic causation of behaviour, she quotes Bertrand Russell, who says nobody could possibly have sufficient information to make behaviour predictable in practice. This leads one to think in terms of the explanatory power of her theory, if, as she states, behaviour is so diverse determinism, 'remains something highly theoretical and remote.'[61] Under her own admission her biological determinism must be somewhat limited in what it can tell us of real value about human behaviour.[62] For instance R.M.Lockley, in *The Private Life of a Rabbit*, talks in much detail about rabbits' social habits and their relations with each other. But to a rabbit these statements may seem trite and banal, in that they might not say anything of any profound significance about their behaviour. Midgley may be guilty of the same misunderstanding.

In order to present a sound basis for her theories, Midgley has to reconsider the biological standpoint she has taken and recognize that it is not value-free and objective in its approach to animals and can only be less so for humans. The ethologists bring to their experiments their own preconceptions and categories which have such findings to be taken into consideration when analysing their findings concerning animals, as such findings cannot be taken as concrete fact just because the subject under analysis is natural science. It does not mean that the pitfalls of avoiding value judgments have been automatically circumvented. If this form of sociobiology is removed from Midgley's analysis, then, as far as the determining genetic factors are concerned, the explanatory power is considerably reduced. If it can be argued that the sociobiological basis of Midgley's theory is untenable, then the rest of her postulations on human nature fall by the wayside. The explanatory power is so reduced that it becomes meaningless.

Conclusion

In summing up the debate on biological determinism, some wider criticisms of sociobiology should be mentioned, for instance its legitimization of the *status quo* and the incorporation of a reductionist and mechanistic framework are basic flaws in this approach. When looking at the conclusions of such biologists as Symons it is obvious that they take the social arrangements of their society as inevitable because they are reflections of some inherent characteristics. They see the causation process running one way from biology to society, failing to take into account the criticisms that could lead to their position being regarded in the same ideological light as other less traditionally scientific epistemologies. The critical role the individual organism plays in this analysis is implied in the reductionist framework which necessitates the consideration of the smallest possible unit in preference to any complex construct such as society.

The mechanistic and reductionist framework that sociobiology builds on precludes any analysis that tries to incorporate the individual organism into a more dynamic and synthetic approach. Under the mechanistic conception, the individual organism is a solitary part of a machine, one cog among many, not influenced nor influencing the resulting structure. The reductionist conception then takes this mechanistic whole and studies it by reducing it to its component parts and, as the whole or structure has no place in this analysis, it is a mere abstraction from individual elements. The divising of a close link between an organism and its environment is an example of a method of explanation that is precluded by this mechanistic and reductionist framework. This link is impossible to establish using mechanistic terms of analysis. The complex web of interdependencies and transactions between an organism and its environment have no place in a reductionist formulation where the effects of the environment would have been reduced away and the solitary organism left for study. The mechanistic approach would deny that the environment could radically influence an organism. This would follow from the notion of the organism constituted as a separate machine-like cog, and one that could function similarly anywhere. Thus putting the environment in a secondary role. The converse position is positing the environment and the organism as inseparable entities. As Lewis Thomas said when looking at the indiscernibility of the boundaries between the organism and its environment:

> There they are, moving about in my cytoplasm.... They are much less closely related to me than each other and to the free-living bacteria out under the hill. They feel like strangers, but the thought comes that the same creatures, precisely the same, are out there in the cells of seagulls, and whales, and dune grass, and seaweed, and hermit crabs, and further inland in the leaves of the beach in my backyard, and in the family of skunks beneath the back fence, and even in that fly on the window. Through them, I am connected: I have close relatives, once removed, all over the place.[63]

In this way sociobiology has two major problems: first, its mechanistic and reductionist framework make it impossible to incorporate any sense of unity when considering the biological world. It is divided up into small, distinct parts, each examined on their own, without reference to other elements, which might have an inextricable relationship to the object under study. Second, with this conceptual framework, sociobiology limits itself to merely an explanation of the individual. It is unable to incorporate entities such as the environment, society or even family units into its analysis. For example, if biology were to advance sufficiently so that it could predict the characteristics of an individual organism by reference to its genotypic make-up, then, in sociobiological terms, the explanation process would be complete. This prediction, however, would have to be made by postulating a given environment which would be incapable of analysis under reductionist methodology. Even in a situation where sociobiology could predict individual behaviour, it would still be a limited theory as any group activity would remain an enigma. In reality, sociobiology cannot even predict individual behaviour and it probably never will, as it makes arbitrary categorical distinctions between organisms and environment. It is analogous to trying to understand the human being by merely examining the hand. Without the body and the intellect, the hand has no meaning. As with Midgley, maybe the sociobiologist's are in danger of producing accounts of human nature with all the rigours of science but with none of the corresponding rigours of meaning.

Notes

1. Quoted in: Rose,S. Lewontin,R.C. Kamin, L., *Not in Our Genes*, Penguin Books, Harmondsworth, 1985. p6.
2. See Barash,D.P., *Sociobiology and Behavior*, Elsevier, New York, 1977. Barash warns of ignoring innate biological factors, in the interest of species survival.
3. Ibid. pix.
4. It is the conception of sociobiology as a scientific theory that leads some theorists to claim that it is not a form of biological determinism, see. Dawkins, R., *The Extended Phenotype*, Freeman, Oxford, 1982.
5. Wilson, E.O., *Sociobiology: The New Synthesis*, Belknap, Cambridge Mass, 1975. p4.
6. Ibid. p3.
7. Rose et al. p236.
8. Dawkins, R., *The Selfish Gene*, Oxford University Press, Oxford, 1976. p158.
9. Barash. p293.
10. Symons, D., *The Evolution of Human Sexuality*, Oxford University Press, 1979.
11. Ibid. p261.
12. For other exponents of this view see. Kinsey,A.C. et al., *Sexual Behaviour in the Human Female*, Saunders, Philadelphia, 1953. and their, *Sexual Behaviour in the Human Male*, Saunders, Philadelphia, 1948.
13. Symons. p171.
14. See Griffiths,S., *Pornography*, Women's Press, London, 1981. and Dworkin,A., *Pornography: Men possessing women*, Women's Press, London, 1981. For a feminist view of pornography that is an alternative to that expressed by Symons.
15. Symons. p180.

16 Wilson,E.O., 'Human Decency is Animal', *New York Times Magazine*, October 1975.

17 See Midgley,M., *Beast & Man: the roots of human nature*, Methuen, London, 1979. For an example of this viewpoint.

18 Midgley,M., 'Rival Fatalism', in Montagu,A (ed.), *Sociobiology Examined*, Oxford University Press, Oxford, 1980. p22.

19 Bleier,R., *Science and Gender*.

20 Barash *Opcit*

21 See. Sahlins,M., *The Use and Abuse of Biology*, University of Michigan, Ann Arbor, 1976. and 'Sociobiology, a New Biological Determinism', in Science for the People Collective, *Biology as a Social Weapon*, Minneapolis, Burgess, 1977.

22 Rose *et al*, p247.

23 Ibid. p248.

24 See Wilson,E.O., *On Human Nature*, Harvard University Press, Cambridge Mas, 1978.

25 Wilson. *Sociobiology: The New Synthesis*, p575.

26 Singer,P., *The Expanding Circle*, Farrar, Strans & Giroux, New York, 1981. p5.

27 Ibid. p63.

28 Ibid. p64.

29 Ibid. p68.

30 Ibid. p69.

31 Ibid. p70.

32 Ibid. p72.

33 Ibid. p73.

34 Ibid. p76-7.

35 Ibid. p79.

36 Ibid. p72.

37 Ibid. p84.

38 See the previous section of this paper for an elucidation of the scientific status of sociobiology and animal human comparisons.

39 For an illuminating account of the distinction see. Lee,K., *A New Basis for Moral Philosophy*, Routledge & Kegan Paul, London, 1985. and the following proponents

of naturalism. Searle,J, 'How to Derive "Ought" from "Is"', *Philosophical Review,* 1964. Bhaskar,R., *The Possibility of Naturalism,* Harvester Press, Sussex, 1979. A comprehensive collection on the issue is, Hudson,W., (ed.) *The Is/Ought Question,* Macmillan, London, 1979. For a critique of Midgley's naturalism see. Cottingham,J, 'Neo- Naturalism and its Pitfalls', *Philosophy,* 58, 1983.

40 Rose *et al*, p237.

41 Wilson, *Sociobiology: The New Synthesis*, p575.

42 Midgley,M., *Wisdom, Information, & Wonder: What is knowledge for?*, Routledge, London, 1989. p72.

43 See Chapter 9 in *Wisdom, Information & Wonder.*

44 For a detailed discussion the fact and values see, *Wisdom, Information, & Wonder,* for Midgley's account. For a critique of her position see, Cottingham,J, 'Neo-Naturalism and its Pitfalls'.

45 Midgley,M, 'Gene-Juggling', in *Sociobiology Examined,* Montagu,A (ed.) Oxford University Press, Oxford, 1980. p132.

46 Midgley,M., *Beast & Man*:

47 Ibid. p19.

48 Ibid. p21-22.

49 See. p27.

50 Ibid. p16.

51 Ibid. p17.

52 Ibid. p27.

53 Ibid. p26.

54 Ibid. p24.

55 Ibid. p347.

56 Noske,B., *Humans and Other Animals,* Pluto Press, London, 1989.

57 Midgley *op cit.* p350.

58 Ibid. p58.

59 Ibid. see. p67.

60 Ibid. p64.

61 Ibid.

62 For a useful critique of the limited explanatory power of ethology, see, Barnett,S,A., *Biology and Freedom,* Cambridge University Press, Cambridge, 1988. Barnett

states: 'The various concepts of instinct at best only name phenomena that are better described in plain words. The notion that we are driven by mysterious inner forces does not explain but only obscures human motivation.' p55.

63 Thomas,L., *The Lives of a Cell*, Allen Lane, London, 1980. p86.

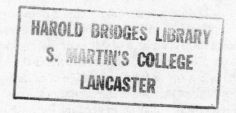

9 The 'evolutionary paradigm' and constructional biology

Brian Goodwin, Gerry Webster and Joseph Wayne Smith

Introduction

An adequate theoretical biology must be able to address the problem of biological form, and consequently to address the dual problems of morphogenesis and taxonomy. It is argued here that the 'evolutionary paradigm' of modern biology fails in both respects. This failure is in part due to the strong genocentric assumptions of this tradition. We advance empirical, theoretical and philosophical criticisms of this tradition, and propose that a step towards dealing with the problem of biological form lies in the adoption of an alternative ontology of the organism: a field conception of organisms.

We suggest that one of the central problems for a theoretical biology is that of biological form; that is, to account for how it is that organisms of a specific morphology are generated (the problem of morphogenesis), and to supply an objective classification of the diversity of natural forms found in nature based upon real relationships between organisms of different form (the problem of taxonomy). It is the thesis of this chapter that 'Darwinism' and the 'neo-Darwinist synthesis' are manifestations of, and have reinforced, a systematic conceptual and metaphysical framework-an 'evolutionary paradigm'-with which it has proved impossible to provide satisfactory answers to both the problem of morphogenesis and the problem of taxonomy.

We shall seek an alternative, as two of us have already proposed elsewhere,[1] consonant with the tradition of 'Rational Morphology' and its concern with the 'laws of form'. Since this tradition was eclipsed by Darwinism, we shall begin this paper by an historically orientated consideration of the Darwinian and Rational Morphological positions on the questions of the relations of natural forms. We then turn to the task of arguing for the inadequacy of the 'evolutionary paradigm', as the basis for a satisfactory and comprehensive theoretical biology. The paper is concluded by a discussion of what we take to be the first step towards a more satisfactory position.

Darwinism and neo-darwinism on form and development

Both Darwinism and neo-Darwinism attempt to formulate the problem of morphology and the problem of taxonomy within an historically based view of biological processes. Morphology, for at least some biologists working within the 'evolutionary paradigm',[2] is taken to be result of the self-assembly of gene products into higher biostructures, whilst taxonomy for both Darwinism and neo-Darwinism, stands as little more than a genealogy of forms, themselves formed by the natural selection of phenotypes in randomly varying environments.

Darwin in *The Origin of Species*,[3] employed two distinct yet related principles to explain the similarities and the differences systemized in taxonomy. Similarity is explained in terms of 'propinquity of descent' from a common ancestor, that is, 'inheritance' provides a satisfactory explanation of morphological aspects of organisms which remain invariant in lineages. Differences are explained as being a result of natural selection from an original continuum of individuals. We shall analyse the Darwinian position in confrontation with Rational Morphology over these two issues in more detail.

The problem of the sameness of individual organisms is the problem of formulating the criteria used for identifying the morphological elements of which these individuals are comprised and assigning them to a given species of 'elements'. To rely solely upon shared physical qualities leads to difficulties when taxonomic comparisons are made between individual organisms at levels higher than, say, the species. For example, is it reasonable to give the same name to various bones of the vertebrate limbs, even if they are physically quite different in size, perhaps shape, strength and other qualities?

Confronted by this problem the Rational Morphologist E. Geoffrey St. Hilaire proposed his 'principle of connections'. Criteria of identify of parts for Geoffrey St. Hilaire required formulation with respect to the spatial relationships between the parts. Organisms were viewed as systems, the parts as elements of such systems being identified and characterized by purely relational criteria of the place or position occupied in the system. Sameness of parts could thus be based on the criterion of the sameness of place in the system, rather than on sameness of significant physical qualities. Consequently, homological relations are viewed as relations of relations, rather than relations between substantive things.[4]

We have with Geoffrey St. Hilaire a science of pure morphology. With this conception of organisms, it is meaningless to speak of 'duplicated elements' or 'missing elements', or indeed to consider, from a comparative point of view, an organism as a loose assemblage of parts, which are atomistically open to natural selection. This is an insight which, simple as it may seem, has been increasingly lost to the neo-Darwinian tradition culminating in the recent views of sociobiologists.[5] It was an insight also lacking, as we shall now argue, from the very first edition of *The Origin of Species*.[3]

For Darwin, homology is a relation of spatially arranged individual things, and sameness is explained in terms of these individuals. All vertebrate limbs for example, include the 'same bones, in the same relative positions [. . .] Hence the same names can be given to the homologous bones in widely different animals' (ibid., pp.415-416). Unity of type is, in short, explained by unity of descent.

Organisms are the subjects of evolutionary history, effectively enduring and suffering adaptive changes in their qualities in response to the dual process of inheritance and the natural selection of the fit. Such individuals, and their elements as well, comprise an aggregate rather than a system. This can be seen from an analysis of Darwin's concept of inheritance (i.e. descent) which is simply that of a causal relation between the individual elements of the parent and offspring organisms.[4] The only known cause of the similarity of organic beings in Darwin's opinion is 'propinquity of descent' and he claims: 'On this view of descent with modification, all the great facts in morphology became intelligible' (ibid., p.433). The attempt, however, to push the problem of form into that of the phenomena of ancestral forms does nothing to resolve the problem of the production of such forms. For the question must immediately arise about the production of ancestral forms. This circularity is similar to that seen by Stallo[6] in the kinetic-molecular theory's explanation of the elasticity of gases by reference to the elasticity of their constituents. The point remains that the synchronic phenomena of natural forms and their relations are not explained by the common ancestry of similar forms. It is thus not surprising that neo-Darwinists such as de Beer[7] find that homology is an unsolved problem since it has proved impossible to discover any material basis for the continuity (direct or indirect) of individual elements. The difficulty, in short, must lie with the neo-Darwinist ontology of the organism.

With regard to the problem of difference, Darwin and the neo-Darwinists after him, have inferred a gradual, historical differentiation of a continuum directly from the facts of individual variation and the practical difficulties encountered in constructing empirical taxonomies, with a consequent denial of the possibility of achieving the taxonomic goal of the discovery of 'laws of form' other than the universal law of functional adaptation.[4] Darwin states with regard to the concept of species, that it is 'arbitrarily given for the sake of convenience of a set of individuals closely resembling each other, and that it does not essentially differ from the term variety, which is given to less distinct and more fluctuating forms. The term variety, again, in comparison with mere individual differences, is also applied arbitrarily and for mere convenience sake'.[3:108] Hull[8,9,10] and Mayr,[11] also viewed Darwinian theory as discrediting essentialist modes of thought which attempt to discover 'laws of form'. The 'form' of Rational Morphology is an illusion, as firstly, only the individuals of a population have reality and secondly Darwinian gradualism is taken to make it impossible for the essentialist to say exactly where one species ends and another begins.

On this view then, species-names are not the names of natural kinds, but are the names of specific individuals, so that no biological law-like statements are applicable to them. Note however, that if Darwinian gradualism does imply that it is impossible to say exactly where one species ends and another begins, this is not merely an argument against the essentialism of the Rational Morphologists but constitutes an effective argument against the species-nominalism of Hull. For if we follow the Quinean maxim that 'there is no entity without identity' a widely accepted maxim then the absence of any theoretical possibility of identifying and individuating species is a good reason to deny that species exist even as individuals.

The tendency within the 'evolutionary paradigm' to reject any explanatory principles of morphological invariance is clearly seen in recent theories such as Wolpert's[12] theory of positional information. Here, organisms are taken to be of an infinitely-modifiable form (hence always open to the opportunistic advances of natural selection) because the only constraint on structure, other than that specified by the genome, is common to all (multicellular) organisms. Thus regularity of the form of organisms is not regarded as a phenomenon requiring explanation, and stands as accidental rather than systematic in nature, the product of chance rather than laws operating within organisms.

The satisfactoriness of the Darwinian and neo-Darwinian responses to the problems of morphology and taxonomy hinges upon the theory of inheritance. Now a significant theoretical innovation in the theory of inheritance occurred with Weismann,[13] where inheritance was thought of in terms of the control of growth and development rather than in the terms of direct material continuity between observable individual elements of the patterns of parents and offspring.

Weismann divided organisms into two distinct parts, the germ plasm and the somatoplasm. The former was localised in the cell nucleus and actively directed the course of growth and development and stood as a repository of all the specific causes of heritable form. The germ plasm consisted of 'active' particulate units ('determinates') which stood in a specific causal relation to particular parts of the manifest pattern of the organism, the relationship between determinates and heritable parts being 1:1. Given the classical empiricist view that constant effects imply constant causes, offspring resemble parents because they are both the effects of identical processes of growth and development, regulated by the materially continuous nature of the germ plasm.

The somatoplasm, on the other hand, is passive and plays no generative role in the process of inheritance. There is no path of 'informational' return from adult to germ plasm so that all inheritance is via the germ plasm. This aspect of Weismann's doctrine made a strong theoretical case against the postulation of Lamarckian-style hereditary mechanisms which in themselves raised grave problems for the theory of natural selection. If the environment directly induced inheritable changes of orgasmic form and adaptive traits, then natural selection is a superfluous force.

It is evident that no competent geneticist today accepts that the relationship between the material substratum of heredity and the phenotype of the organism, is 1:1; polygenes and pleiotropic gene action are well known. Nevertheless, embodied in the appropriately named 'central dogma of molecular biology',[14] where reverse-transcription and reverse-translation processes are denied, is a contemporary restatement of Weismann's doctrine of the insulation of the germ plasm. However, as is well known, an enzyme responsible for reverse-transcription, RNA-dependent DNA polymerase ('reverse-transcriptase') has been detected in normal animal cells[14] and so this scheme has undergone revision. There has also been suggestion of reverse-translation mechanisms, such as in Mekler's reverse translation hypothesis, which does not involve the denaturing and exact decoding of proteins.[15]

Regardless of such models, a significant amount of biological evidence exists to indicate that reciprocal interrelations exist within organisms at all levels, including the molecular. Nucleic acid depends upon proteins for replication and specific proteins bind to nucleic acids to alter the rate of transcription. None of this should be at all surprising to biologists and biophilosophers, given the general acceptance of the Jacob-Monod operon model of the gene[16] where regulatory metabolites produced by either other operons or by substances such as hormones[17] in combining with repressors, regulate the behaviour of structural genes. However, against this well known picture, it has been recently argued in response to us, that Weismann and the supporters of the 'central dogma' are not concerned with matter or energy but information: 'Weismann was saying that information (not material or energy) flows among the germ line, from germ line to soma, but not from soma to germ line'.[18:50] We shall return to this later, pursuing first one final issue centred around the contemporary neo-Darwinist view of organisms.

The final element of Weismann's scheme is that the germ plasm is the sole vehicle of inheritance, causally responsible for the somatoplasm. This assumption is also part of the neo-Darwinist position and is explicit in the genetic calculus of the sociobiologists. Here the genotype is taken to be the sufficient cause of the phenotype, and the metaphors used to explicate this causal relation are invariably taken from cybernetics. For Wolpert,[12] genetic information contains a set of 'instructions' for the construction of an organism. The metaphor varies and the concept of a 'genetic programme' is also used, based upon an analogy with computer software.[11]

We shall attempt to challenge the neo-Darwinist view of organisms at each of the above specified points, and thus advance a case against 'genocentric biology'. Such a position is, it is maintained, open to empirical, theoretical and philosophical objections, which we now detail.

The case against the 'evolutionary paradigm'

There is no doubt that the advance of molecular biology in the second half of the twentieth century has been impressive. Further, we do not deny the frequently very significant impacts which protein synthesis malfunctions have upon structure and behaviour; this is clearly indicated by various genetic defects. Nothing of philosophical significance follows from this admission, however. Just as a loose bolt in a complex mechanism such as a computer may, through multiplied effects, have a significant impact upon its behaviour, so too may genetic and chromosome defects have significant impacts upon behaviour. It does not follow from this that differentiation, development and resultant behaviour can be solely understood from a knowledge of protein synthesis plus other biochemical variables and their mutual interactions. In any case, genetic defects hardly illustrate how 'genetic information' or 'instructions' produce morphology, and we find that no satisfactorily detailed account of morphogenesis by genocentric biology exists. There is good reason for this we will argue: its central metaphors are incoherent.

In reply it may be argued that frequently in science the full details and knowledge of specific mechanisms are lacking and must in due course be filled in by future scientific research. Thus, it would be maintained, the absence of a genocentric and physico-chemical theory of morphology would hardly be a cogent reason for scepticism about genocentric biology itself. Rather, morphology is on the agenda of research.

This argument is fallacious, and is little more than an expression of happy faith. Granted: morphology is on the agenda of research-nothing follows about whether a satisfactory genocentric theory of morphogenesis will be forthcoming.

In the absence of a specific theory, appeal is usually made to metaphors, most notably from information theory and computer science. We find such appeals misleading. Let us take up the appeal to information theory first.

Information theory is about communication systems. These essentially comprise a source of messages, a communication channel and a receiver of messages. The amount of information in a given message H is given by $H = \log_2$

(p1/p2), where p1 is the probability at the receiver with respect to an event after the information has been received, and p2 is the probability before the information is received. True, we speak here of the 'amount of information', just as we speak of the amount of matter, but 'information' is a purely relational phenomenon, being nothing more than a probabilistic relationship between events in a communication channel. It does not have causal import on particulars in the world, because it is not a causal relation. We have argued elsewhere[1] that the proper way to use this concept in an ontogenetic context is in relation to the actualisation or stabilisation of one of the possible processes (transformations) available to an organism at particular times in its development. Thus we can speak of the information associated with the influence of gene products in selecting among the possible transformations or trajectories which a developing organism can follow, but we cannot say that this information creates or generates the possible transformations. What is absent from an account of morphogenesis based upon the concept of information which 'flows along the germ line',[18] is the origin and nature of the developmental processes among which selection is being made by the genetic information. Without the definition of this set, the use of information theoretic ideas in development is meaningless, so that Maynard Smith's[18] interpretation of Weismann is unsatisfactory and remains, at least, metaphorical.

When one metaphor fails, try another, so let us consider the metaphor of the 'genetic programme'. Logically and metaphysically, computer programmes cannot be causal agents as such, but at best can only describe processes in mathematical form. The explanatory power of such programmes is debatable, although they may have predicative power.

Biologically speaking the genetic programme is considered to be a set of instructions or algorithms, written in the DNA, which determines both the set of possible processes which an organism can follow at any stage in its development and the decision procedures for selecting those processes which lead to an organism of defined morphology.[19] Thus it is a more comprehensive concept than information, and exercises a very powerful influence on the biologist's imagination as a potentially adequate, though as yet incompletely worked out, framework for explaining generative processes (reproduction and regeneration). Let us make it clear that what we are questioning here is the idea that a genetic programme as described above could contain a sufficient set of instructions for development, not the more general possibility that development can be stimulated by means of computer programmes, a quite different proposition to which we subscribe.

As far as we know, no-one has actually defined a genetic programme for embryogenesis. However, Monod's[2] account of genetic control processes and self-assembly of gene products might be taken by some as an example of such a theory. The empirical evidence against this comes from cases where it has been established that gene products are not in themselves sufficient to determine the form or morphology of the structures into which these gene products are assembled. Thus Oosawa et al[20] have shown that the protein flagellin from wild-type flagella of *Salmonella* can form two morphologically-distinct flagella forms, wavy and curly, depending upon the 'seeds' or nucleation centres that are added to a solution of flagellin to initiate the assembly process. Sonneborn[21] has argued convincingly that this type of nucleation process underlies the formation of cortical structures in ciliate protozoa, in which cellular morphology is inherited in a manner that is independent of DNA. There is nothing in the least surprising about such phenomena, which simply show that structural polymorphism occurs in macromolecular 'crystals' just as it does in inorganic crystal formation; composition does not determine form. What the genotype does is to define the potential macromolecular composition of the organism, but it is, in general, insufficient to define its form.

Evidence likewise exists that, at a higher level of self-assembly, the same cells can produce different structures in higher organisms. Thus cells whose normal fate is to form the limb in chick embryos can be spatially scrambled after which they form recognisably limb-like but abnormal structures.[22] Similarly, a variety of abnormal limb forms arise from a reordering of spatial relations in regenerating limbs of the axolotl, again pointing to structural polymorphism in organisms despite constancy of genome and of the external environment.

The absence of a one-to-one relationship between genotype and phenotype is shown by the occurrence of morphologically normal *Drosophila*, possessing mutant genes which 'cause' abnormal morphology,[23] some individuals having mutant morphology on one side of the body and being normal on the other. Thus one genotype is consistent with many phenotypes, even with the environment constant. Recently it has been shown that the same structure (thoracic leg) can be generated by the action of different genes in *Drosophila*[24] and the same morphology results from different gene products in *Tetrahymena*.[25] Thus a particular pattern of gene activity is neither necessary nor sufficient to generate a particular morphology. Since the whole point of the computer metaphor is to identify the generators of organismic morphology with defined patterns of gene action, we see that this proposition is also untenable and must be abandoned. What lies behind both the information theoretic and the computer programme analogies, applied to 'gene information' or 'instructions' in the DNA, is the 'root metaphor'[26] of a central directing agency which was first formulated by Weismann[13] in his hypothesis of the germ plasm an entity with a 'highly complex structure' which has 'the power of developing into a complex organism'. We have argued elsewhere[1] that, historically speaking, this may be understood as a materialised form of the concept of a directing Spirit, Soul, or Idea which nineteenth century German holists had used to explain the form of organisms. The matter-spirit dualism of the holists became the somatoplasm-germ dualism of Weismann, and then the energy-information or process-instruction dualism embodied in the information theoretical and computer programme metaphors. An alternative root image is required, and in the final section of this paper we would like to make some preliminary suggestions as to what this might be.

The field properties of organisms

What is evidently absent from the conception of organisms employed in the 'evolutionary paradigm' is any adequate account of the relational spatial order which underlies both reproduction and regeneration, the capacity of parts of organisms to transform into wholes. The first person to emphasize this property was Hans Driesch[27,28] who observed that each of the two cells (blastomeres) produced after the first cleavage of the fertilised sea urchin egg developed into a completely normal, though half-sized pluteus larva, which then grew to normal size. This property of 'regulation' is now recognized as a universal feature of developing organisms and it reveals that organisms have field properties. A field is a spatial domain with the property that every part takes on a state determined by that of neighbouring parts, so that the whole has a specific relational structure or order. Perturbations of fields, such as removal, addition, or rearrangement of parts, results in a restoration of overall order so that an organised whole is produced.[29,30] It is this property of organisms which underlies both reproduction and regeneration, for the essential feature of these processes is that from a part a whole is generated. This part may be a single cell, the oocyte, as in sexual generation; it may be a group of cells which forms a bud as in asexual generation in plants or in animals such as hydra: or it may be a single cell which is the whole organism, as in the unicellulars (in which case 'part' means its upper limit, which is the whole, so that one whole organism generates another). Similarly, the part which restores the whole in regeneration may be either many cells cooperating and interacting to restore the missing part, as in axial regeneration in planaria or limb regeneration in insects and urodeles (tailed amphibians such as newts); or it may be a fragment of a cell which restores the whole organism, as in the unicellular ciliate protozoa.

What we see from these examples is that the cell is not the basic biological entity underlying the generative (reproductive) process, and certainly it is a mistake to identify this process with a particular category of cells, the germ cells containing the germ plasm, as Weismann argued; nor is generation to be understood in terms of the properties of DNA as in the contemporary view. Rather, generation is to be understood as a process which arises from the field properties of the living state, the inherited variety of form observed in organisms arising from the inheritance of particular influence such as specific gene products or nucleation centres, acting as parameters contributing to the specification of initial and boundary values of the *field* equations.

The exact nature of the generative field equations is an active area of research[31,32,33,34,35] and it seems likely that different types of fields arising from diffusion and reaction, from visco-elastic properties of cells, from electrical potentials and ion flows, and so on, may contribute in different ways to different morphogenetic processes. A more detailed understanding of these processes should lead to greater insight into the constraints on developmental transformations (sequences of field solutions) which result in the hierarchical relationships and the discrete distribution of different morphologies which underlie taxonomy. Thus developmental processes become the basis for a rational taxonomy of natural kinds and evolution may then be understood as the selection of a stable subset of the possible forms of organisms by the action of environmental contingencies. The emphasis of such a view is thus on the generative origins and transformational relationships of organisms, so that biology becomes grounded in a generative theory of transformations.

This is an ambitious claim and one which we can hardly systematically justify in the space available in this paper. It is thus an expression of an alternative research programme to that of the prevailing 'evolutionary paradigm', with a shifting away from the utilitarian, empirical and historical preoccupations of Darwinism and neo-Darwinism and a turn toward the consideration of ordered, synchronic and harmonious aspects of living organisms. Thus far, however, our terms are indeed quite metaphorical. Note however that our criticism of the 'evolutionary paradigm' was not its use of metaphors, but its use of inappropriate metaphors. The 'Structuralist' or 'Constructional' view of biology is at least capable of precise elaboration.

Conclusion

Recent traditions in the philosophy of science including Kuhn[36] Feyerabend,[37] Lakatos[38] and Laudan[39] are united in claiming that no paradigm research programme/research tradition, is ever rejected until a satisfactory alternative is presented. Our claim here is that the 'evolutionary paradigm' does not enable a satisfactory theoretical biology to be constructed. Further, this claim has been made since 1859, and by both the Rational Morphologists (see Russell,[40]) and later by Bateson[41] and Driesch[28] among others. Their criticisms have not in our opinion been satisfactorily dealt with by either Darwin or the neo-Darwinists. Further, we maintain, as we have previously argued[1] (Webster,[4] that the Rational Morphologists came somewhat closer to presenting a satisfactory basis for a unified theoretical biology. Hence we disagree with received post-empiricist metascientific wisdom: at least one research tradition has been rejected before a satisfactory alternative was presented. That is the research tradition of Rational Morphology.

Notes

1 Webster, G. and B.C. Goodwin, 'The Origin of the Species: A Structuralist Approach', *Journal of Social and Biological Structures*, 1982, 5:15-47

2 Monod, J., *Chance and Necessity*, London, Collins, 1972

3 Darwin, C., *On the Origin of the Species By Means of Natural Selection*, London, John Murray, 1859.

4 Webster, G.C. 'The Relations of Natural Forms', in *Beyond Neo-Darwinism*, eds. M-W. Ho and P.T. Saunders, London, Academic Press, 1984

5 Gould, S.J. and R.C. Lewontin, 'The Spandrels of San Marco and the Panglossian Paradigm: A Critique of the Adaptionist Programme', in *Proceedings of the Royal Society of London*, Series B, 1979, 205:581-598

6 Stallo, J., *Concepts and Theories in Modern Physics*, Cambridge, Mass., Harvard University Press, 1960

7 De Beer, G., *Homology: An Unsolved Problem*, Oxford, OUP, 1971.

8 Hull, D.L., 'The Effects of Essentialism on Taxonomy: 2000 Years of Stasis', *British Journal for the Philosophy of Science*, 1965, 15:314-326

9 Hull, D.L., 'Are Species Really Individuals?', *Systematic Zoology*, 1976, 25:174-191

10 Hull, D.L. 'A Matter of Individuality', *Philosophy of Science*, 1978, 45:335-360

11. Mayr, E., 'Behaviour Programmes and Evolutionary Strategies', *American Scientist*, 1974, 62:650-659

12 Wolpert, L. 'Positional Information and Pattern Formation', *Current Topics in Developmental Biology*, 1971, 6:183-224

13 Weismann, A., 'The Continuity of the Germ-Plasm as the Foundation of a Theory of Heredity', (1885). Reprinted in J.A. Moore, ed. *Readings in Heredity and Development*, New York, OUP, 1972

14 Watson, J.D., *Molecular Biology of the Gene*, 3rd edition, California, Menlo Park, W.J. Benjamin Inc. 1977

15 Cook, N.D., 'The Case For Reverse Translation', *Journal of Theoretical Biology*, 1977, 64:113-135

16 Jacob, F. and J. Monod, 'On the Regulation of Gene Activity', in *Cold Spring Harbor Symposia On Quantitative Biology*, 1961, 26:193-209

17 Korner, A., 'Hormone Control of Protein Synthesis', *Proceedings of the Royal Society of London*, 1970, Series B, 176:287-290

18 Maynard Smith, J., 'Commentary on Webster and Goodwin's "The Origin of Species: A Structuralist Approach"', *Journal of Social and Biological Structures*, 1982, 5:49-51

19 Jacob, F., *The Logic of Life: A History of Heredity*, trans. B.E.Spillman, New York, Pantheon Books, 1973

20 Oosawa, F., Kasai, M., Hatano, S. and S. Asakwa in *Principles of Biomolecular Organisation*, eds. G.E.W. Wolstenholme and M. O'Connor, Boston, Mass., Little Brown and Co., 1966, 272-303

21 Sonneborn, T.M.C., 'Gene Action in Development', *Proceedings of the Royal Society of London, 1970*, Series B, 347-366

22 Zwilling, E., 'Development of Fragmented and Dissociated Limb Bud Mesoderm', *Developmental Biology*, 1964, 9:20-37

23 Postlethwaite, J.H., and H.A. Schneiderman, 'Pattern Formation and Determination in the Antenna of the Homoeotic Mutant Antennapedia in Drosophila Melanogaster', *Developmental Biology*, 1971, 25:606-640

24 Morata, G., and S. Kerridge, 'The Role of Position in Determining Homoeotic Gene Function in Drosophila, *Nature*, 1982, 300: 191-192

25 Williams, N.E., 'An Apparent Disjunction Between the Evolution of Form and Substance in the Genus Tetrahymena', *Evolution*, 1984, 38 (1): 25-33

26 Pepper, S.C., *World Hypotheses: A Study in Evidence*, Berkeley and Los Angeles, University of California Press, 1961.

27 Driesch, H., 'The Potency of the First Two Cleavage Cells in the Development of Echnioderms', (1892) in B.H.Willier and J.M. Oppenheimer, eds. *Foundations of Experimental Embryology*, Englewood Cliffs, Prentice Hall, 1964, 38-50

28 Driesch, H., *Science and Philosophy of Organism*, London, A. and C. Black, 1908

29 Huxley, J.S., and G.R. de Beer, *The Elements of Experimental Embryology*, Cambridge, C.U.P. 1934

30 Waddington, C.H., *Principles of Embryology*, London, Allen and Unwin, 1956

31 Murray, J.D., 'A Pre-Pattern Formation Mechanism for Animal Coat Markings', *Journal of Theoretical Biology*, 1981, 88:161-199

32 Meinhardt, H., *Models of Biological Pattern Formation*, London, Academic Press, 1982

33 Odell, G., Oster,G.F. Burnside, B. and P. Alberch, 'The Mechanical Basis of Morphogenesis', *Developmental Biology*, 1981, 85:446-462

34 Goodwin, B.C. and L.E.H. Trainor, 'A Field Description of the Cleavage Process in Embryogenesis', *Journal of Theoretical Biology*, 1980, 85:757-770

35 Goodwin, B.C., 'A Relational or Field Theory of Reproduction and its Evolutionary Consequences', in *Beyond Neo-Darwinism*, eds. M-W Ho and P.T. Saunders, London Academic Press, 1984

36 Kuhn, T.S., *The Structure of Scientific Revolutions*, 2nd enlarged edition, Chicago, Chicago University Press, 1970

37 Feyerabend, P.K., *Against Method*, London, New Left Books, 1975

38 Lakatos, I., 'Falsification and the Methodology of Scientific Research Programmes', in Lakatos, I. and A. Musgrave, eds. *Criticism and the Growth of Knowledge*, Cambridge, C.U.P., 1970

39 Laudan, L., *Progress and its Problems: Towards a Theory of Scientific Growth*, Berkeley, University of California Press, 1977

40 Russell, E.C., *Form and Function*, London, John Murray, 1916

41 Bateson, W., *Materials For the Study of Variation*, Cambridge, C.U.P., 1984

10 Challenge of ill health

E.K. Ledermann

Introduction

Patients seek treatment from practitioners who follow either mechanistic-scientific or holistic principles. This paper examines how sufferers actually experience their illness and also how the application of the above principles may affect the way they try to cope with their illness. Those who struggle with an illness meet a challenge, presented by their conscience which imposes an obligation not to let either themselves or others down. Their strength in meeting this challenge may vary, and therapists have to be aware of such variations and provide support when patients are at a low ebb. But more than support may be required. Therapists may have to point to a way ahead of which the sufferers may not be aware. They may have to be convinced that new attitudes must be adopted as the old ones have been related to their illness.

Science and technology

Science and technology provide a fundamental orientation to most modern people. One result of scientific medicine is that individual patients are depersonalized as bearers of diseases and doctors are trained to diagnose and treat diseases specifically. New treatments are evaluated by comparing homogeneous groups of sufferers from their particular ailment, half of whom receive the remedy, the other half receiving a placebo. On the basis of statistical evaluation a decision is reached concerning whether the group receiving the therapy have benefited. This method is obviously necessary for medical progress and scientific accuracy, but in another sense it constitutes a devaluation of the person.

To consider an example: although a hundred patients suffering from rheumatoid arthritis may have identical physical manifestations of their illness, as evident in the X-ray photographs of their joints and in the findings in their blood, as sufferers of the disease they are by no means equal. Each and every one of them experiences his or her complaint in a personal way and has to cope with it individually. Their particular surroundings may play a major part and the support - or lack of it - from relatives, friends, or members of the comunity, will be of great importance to them. But of even greater importance is the power of their own physical and mental constitution; the extent to which their own organisms can withstand the ravages of the disease and how their minds and spirits can cope with the accompanying disability. A scientifically trained doctor would fail his or her patients if he or she only acts as a medical scientist ordering laboratory investigations and prescribing treatment accordingly, without awareness of the vital personal element and meeting the patients as sick persons.

The inevitable specialization of modern scientific medicine has contributed to a decline in personal medical attention. Patients are sent from one department to the next, with each specialist concerned only with one particular part of the body - or with the mind - an overall situation which leaves little or no room for concern for a unique human being.

In the institutional surroundings of scientific medicine patients join their doctors in the impersonal world of science. They expect to be given information in terms of disease labels and doctors in turn are encouraged to explain in lay language the meaning of the scientific terms employed in the description of disease processes.

Although it is right and proper that patients should receive such information and should be told of the significance of the diagnosis, they ought to be aware of the limitations of such understanding. They should, for example, be aware of the ways in which their personalities respond to the troubles which have been diagnosed.

The extraordinary lengths of the impersonal nature of medical technology was recently recorded in a rather jocular report in one of the United Kingdom's quality newspapers. This report concerned the idea of an 'intelligent lavatory'. In about four years, said the report, 'if the Japanese have their way, your WC will provide you with a complete medical check-up. Micro-chips will bear sensing elements which will monitor your waste products, alerting you to malfunctions. When you sit on the intelligent lavatory it will weigh you, it will record your body temperature. It will somehow take your blood-pressure.' There are also plans, says the report, that 'your intelligent lavatory will be "on line" to the local hospital. Suspicious symptoms will be logged by the hospital computer. You will automatically be referred to a doctor. Presumably if you do not wish for medical attention, official persons - perhaps a crack squad of armed plumbers will come around to make sure you get it. The potential for confusion is vast. How will your lavatory know it is you? Caught short, will you have to key in a personal number? Voice recognition is a possibility; but you may think you have better things to do than talk to your sanitary ware'. The report concludes: 'Only the Japanese could have invented this device to make one's most private moments public property'. [1]

The intelligent lavatory may turn out to be a piece of scientific fiction but it symbolizes scientific medicine's tendency to make bodily functions into public property. In this case, however, it also includes mental functions such as anxiety, as these cause abnormal functioning of kidneys and bowels.

Psychological medicine as a scientific discipline inflicts far more harm than physical (bodily) medicine on the human personality. For it is confined to the straitjacket of determinism which denies personal freedom and the responsibility which is its manifestation. Those who accept the world of the sciences of the mind are turned into robots, differentiated according to the dogmas of the various schools: cerebral by the organic psychiatrists, reflexological by the behaviour therapists, libidinal by the psychoanalists and archetypal by the analytical psychologists. True to the spirit of science, each school operates with mechanistic concepts. The sciences, concerned with the diseases of body and mind, occupy public impersonal worlds, the sufferers from bodily and mental complaints occupy personal private worlds. The two worlds meet in doctor's consulting rooms and hospitals.

Mrs Jones' nervous stomach

The consultation with the general practitioner

On an average seven minutes are allocated for a consultation. Patients will have braced themselves for this meeting with their doctors. They will have had to overcome their shyness and hesitation. The time spent in the waiting room has added to their confusion and fears. They are often incoherent.

Doctors take a history to extract clues for the diagnosis of the physical or mental disease and conduct an examination of the body or mind to arrive at the diagnosis from which the specific treatment is to follow. Doctors are frequently disappointed, textbook cases of diabetes, hyper- or hypothyroidism and other treatable conditions are rare. The reason is that, for instance, Mrs Jones' 'nervous stomach' is the result of her domestic troubles, her husband's drinking which makes him violent and her son's drug addiction which has brought him into conflict with the law. Those relevant factors are not likely to emerge in the consultation. Mrs Jones, who is obviously anxious is diagnosed as a psycho-somatic case. The psychological part may, in the doctor's opinion, call for a sedative. He has been warned not to prescribe such drugs for lengthy periods, but the situation at home is not going to change within a week or two. Thus Mrs Jones will require the sedative again and again and will become dependent on it which then constitutes another 'disease' for which the doctor bears some responsibility. The physical part requires physical examination. Mrs Jones is prepared for it, undresses and offers her body for inspection, palpation and ausculation. Although nothing abnormal has been discovered, a stomach medicine is prescribed apart from the sedative. The doctor even prides himself on being holistic, having covered both body and mind in his treatment. If the pain in the stomach persists, a referral to a hospital may be necessary.

The out-patients' department of the hospital

The visit to the hospital is terrifying. The surroundings are unfamiliar, Mrs Jones meets administrators, medical students, nurses, technicians in various laboratories, junior doctors and a senior physician. The results of the numerous investigations are not made known to her. Her doctor will receive a letter in due course. She now has to wait for a further appointment with her doctor, spending anxious days.

If gall stones have been discovered, they, not her domestic troubles, are considered to be responsible for her symptoms. These troubles will be overlooked. An operation is advised.

The in-patients' department

To be away from home is no relief. There is worry about how they are going to cope without her and how she is going to cope with being a patient in a ward with other women. There is the terror of the operation and the bearing of post-operative pain. Her body is being handled by scientific-technological medicine, but it is *her* body which hurts and feels strange and it is *her* mind that receives the effects of the morphia injection. She has difficulties in recognizing herself. A friendly nurse may help a bit, doctors are far too busy and nurses also have little time to talk to patients. Still weak from the ordeal and bewildered she returns to a chaotic home situation. [2]

Non-scientific help

Mrs Jones' 'nervous stomach' cannot be cured by scientific medicine, even after the gall bladder operation. The 'nervous stomach' has to be endured, but this is not a passive situation, rather it entails the activity of moral freedom. This means that Mrs Jones has, like other people, the freedom to act according to her conscience which tells her how she ought to behave. She would feel guilty if she let herself down and did not make a supreme effort to cope with the situation at home and with the abdominal discomfort.

Her life situation varies, sometimes she feels on top of her problems, at other times she is weighed down and feels she cannot cope. On such occasions she needs help to find the strength to manifest her moral freedom. Such help may be provided by a member of her family or by a friend. But they are too emotionally involved in Mrs Jones' situation to sustain her in her efforts. Her doctor - probably her General Practitioner - is the most suitable person.

Doctors can perform such essential services only if they can step out of their scientific medical roles and enter into their patients' crisis situations, understand private worlds from the other person's point of view and help the victims to gain clarity of the whole situation. Such assistance conveys the following message: 'you are being trusted, there is strength in you to cope with your predicament, you have done it before and you will be able to do it again'. The doctor will have to follow the course of the patient's gathering of moral strength so that the person feels sustained all along when climbing out of the depths of despair.

Doctors may have to extend their patient's vision of their situations and may have to draw attention to a moral challenge of which the person was not aware. In that way a disease may appear in a new light. Not only must its impact on life be coped with, but it may also open up new vistas, new attitudes, if people accept the challenge of changing their life style. [3]

Inflamed branchial cysts

A gifted viola player was stopped from playing her instrument because of a painful infection which flared up whenever she pressed her instrument against the side of her neck. The condition was diagnosed as an abnormality that had persisted from her embryonic life. An operation was performed, but failed to cure the condition. A further operation was considered to be too risky, as important nerve tracts may get damaged.

The new approach considered the whole person, especially the possibility of the body effecting a cure if a drastic dietetic change was made to which the body would respond. The new diet consisted mainly of raw fruit and raw vegetables and protein from cheese and nuts. The patient felt weak at the beginning of the treatment, but then her body became used to the new diet and she felt strong. In addition she received several homoeopathic remedies which stimulate the healing power of the body, one to be taken locally, the others taken internally.

This approach was different from the scientific one; the aim was not to influence the inflammation by a specific method which is designed to correct the faulty mechanisms, but to rely on the integrative healing power of nature.

The patient's cooperation in the regime was vital. She realized that by giving up her former diet, which had included cooked food and coffee, in favour of raw food and herb tea, she had taken responsibility for her recovery. She was rewarded. After three and a half weeks on this treatment her neck was much better and she was able to play her viola for up to two and a half hours without any sign of inflammation. After a further three weeks she reported that she was playing her instrument for up to six hours without suffering any discomfort. A follow-up three and a half months later revealed that there had been no recurrence of the trouble in her neck and that she encountered no problems when playing. She had continued with the diet, except for an increase in her consumption of cheese. She had not taken any medicine during the intervening months, as there had been no need for it.

Challenge of a crippled arm

A bodily deformity calls for acceptance and such acceptance transcends the dimension of scientific determinism. It represents a challenge to a person's moral freedom.

A professional single man, aged twenty five, was referred for psychotherapy. He was described as 'very depressed and anxious', worried about the responsibilities of his work because of his depression and lack of sleep and also worried about losing a relationship with a girlfriend because of his 'moodiness'. The patient had seen a psychotherapist before who had related his current symptoms to unhappy experiences in his childhood, specifically to the disharmony of his parents' marriage. The patient had rejected this approach which is based on scientific libidinal theory which denies adult patients their freedom by regressing them to childhood and by making them dependent on therapists who also lose their personal freedom by taking on the role of their patients' childhood parents in the transference situation.

The referring psychotherapist stressed the patient's anger and mentioned his underdeveloped arm, a deformity that dated back to his embryonic life when the nerve supply to the right arm was interfered with, resulting in retardation of growth, the hand being deformed and in adult life no larger than a child's hand. But the psychotherapist did not connect the patient's anger with this deformity, only to 'feeling let down by his parents'. This interpretation would open the road to the use of transference, the therapist taking upon him/her the parental role.

This method was not adopted, rather the patient was faced with the challenge presented by the deformed arm. He freely admitted that because of the small hand he actually felt small as a person and felt unable to make love to a woman as a man. The patient argued that he could not compete with any other man who would not be as handicapped as he was and he admitted that his anger was due to this perceived disadvantage as a lover. The therapist countered this argument by pointing out that the deformity presented a challenge to his freedom and if he met this challenge he would emerge as a stronger and more mature person and in that way would actually have an advantage over any man who did not have to surmount a serious handicap. The patient accepted the strength of this argument. This change of attitude and change of self image could eventually serve him well in a future relationship with a woman.

From emptiness to threefold fullness

A gifted musician of twenty-six consulted a doctor because of severe tension headaches and a general painful muscular tension. She had suffered from having lived under a repressive communist regime which had prevented her family from expressing their opinion. Her parents' house had become delapidated and this had prevented her from inviting friends to the home and from accepting invitations from them. Her mother suffered from depression. Although brilliant at school, she had never been able to credit herself with any success.

As in the previous case, the patient's self image was the target of the treatment and the libidinal historical relations were not considered to be relevant. A reverie provided the clue to the identity crisis.

When asked to close her eyes, relax her body and imagine facing a stage, she pictured a desolate empty scene which filled her with despair. When asked by her therapist to make contact with passers-by, she cried out in painful suffering, caused by their attention! 'I don't want anybody to see my emptiness, I have no idea who I am, it hurts me and makes me feel ashamed that there is only emptiness in me'. During further reveries the stage was dark and filled with frightening figures which she recognized as symbols of her fears.

A turning point in her treatment was reached when a beautiful woman dancer, dressed in white, appeared on the stage which was lit by a lovely blue light. When asked by her therapist to join in the dance and to play suitable music, she expressed that she felt 'unspeakably small, unworthy, dirty, clumsy and ugly'. 'I wanted to stay', she said, 'unnoticed, gazing at the perfect sight'.

Then a second woman dancer appeared which struck her first as frightening and malevolent. She was dressed in black. She began a wild dance. When asked to participate in the dance, the patient pictured herself playing on a piano 'ecstatic, barbaric, devil- like music' and then joining the dancer, moving in a frenzy, passionately, realizing that she had discovered part of herself which included her animal nature which was also symbolized in another reverie by a dark, placenta-like pulsating organ. 'I never have felt so much like my true self', she cried, although she felt that lots of people would disapprove. But she felt confident, happy, relieved, and did not want to hide any more.

When the white dance re-appeared, the figure did not seem so remote any more. This was, she discovered, her 'other self, her soul's aiming at purity, knowledge, clarity, God'.

Finally, a red-clothed female dancer came on the stage inviting her to dance with her. She was now laughing, making funny faces, enjoying herself. This part of herself was identified as her youthful vitality. As a result of this treatment her muscular tension diminished, her musical performance as a flautist gained in depth and expresion. This is how she summed up her gains: 'The basic importance of this three-fold one-ness is that it helped me to discover my true nature. It is is still very hard for me sometimes to free myself from old habits, old patterns of behaviour, but the memory of the three figures removes any difficulties. I now know who I am: black, white and red as the treatment revealed'. She had achieved success in meeting the challenge of her ill health.

Conclusion

Different worlds must be recognized in medical practice. For the patients, their primary world consists of their experience of ill-health. In it, hope and despair, a courageous and a defeatist attitude, anxiety and steadfastness, play a major role, tinged with personal emotions. As Western people, patients accept an orientation, based on medical science which is dominated by the conception of disease. People experience themselves as bearers of disease, some of which such as cancer have a terrifying connotation.

In this world of objective medical science people become fragmented and deprived of their uniqueness and individuality. Their moral freedom is sapped as they experience themselves in the grip of some fateful biological or psychological process.

The world of holistic medicine provides relief from the straitjacket of scientific medicine. Here people's life style is fundamental - which means that they are expected to take responsibility for their health. The emphasis is on the trust in the healing power of nature, a trust in life's mysteries, which is better than a misplaced scientific hubris.

All practitioners have to enter into their patients' private subjective worlds to provide essential support to those who are challenged by ill-health. But more is required: people must be made aware of their potentialities, of the extension of their private worlds which are beyond their current boundaries in which they are able to realize the freedom of their true personalities, recovering from emotional bondage. [4,5]

Notes

1. Mantel Hilary, 'In loo of proper medical advice', *The Independent on Sunday*, September 23, 1990. p.21

2. The depersonalizing elements of scientific-technological medicine are discussed in *Qu'attendez-vous du Medecin?*, Librarie Plon, 1949, which contains contributions from seven French writers.
3. See, for instance, Ledermann, E.K. *Your Health in Your Hands, A Case For Natural Medicine*, Green Books, 1989.
4. The distinction between the subjective private world and the objective public world is the subject of Roubiczek, Paul, *Denken in Gegensätzen*, Vittorio Klostermann, Frankfurt am Main, 1961.
5. Freedom of conscience which is realised when patients meet the challenge of ill health is the subject of Ledermann, E.K., *Mental Health and Human Conscience, The True and the False Self*, Gower Publishers, Aldershot, 1984.

Acknowledgement

I should like to thank Dr. David Lamb for valuable suggestions regarding the formulation of my views.

11 Death: The final frontier

David Lamb

Introduction

Human beings are the only species to manifest moral respect for the dead; the only species to dispose of the dead in a systematic way; the only species to give a meaning to death. Throughout history both religious and secular belief systems have given a meaning to death, whether in terms of the separation of soul and body or simply the ceasing to exist. But whatever meaning was given to death, the fact of death was accepted as an empirical matter. In Western Europe from earliest records, the traditional view was that death occured with the last breath of life. This view is echoed in the Judaeo-Christian and Moslem religions. It was only after circulation was discovered in the 17th century and ausculation introduced in 1819 that absence of heartbeat was seen as a sure sign of death. Until the 1960s criteria for the loss of heartbeat and circulation were seen as unambiguous indicators of death.

The reason why it is necessary to enter into philosophical and scientific inquiries into the meaning and definition of death is that developments in the biomedical sciences, in resuscitation technology, in cardiac transplantation, and even complete artificial mechanical hearts, have created a situation where death is not a momentary event characterised by the simultaneous cessation of all vital systems, and that the mechanism of death is, to a certain extent, independent of heartbeat and circulation.

It is no longer possible to regard death as a brute empirical fact, determined by technical means, and it is therefore both scientifically and morally necessary to formulate a definition of death (primarily a philosophical task) from which criteria and tests for death (primarily an empirical task) can be logically derived.

Concept and criteria

Criteria and tests for death - and arguments about better criteria - are meaningless unless related to some overall concept of what death means. Concepts of death may vary. There is the legal fiction, used in wartime, of 'missing presumed dead', the religious concept of 'separation of soul from the body' and one Brahmin high caste considered a form of social death to be 'marriage outside the social group'. None of these concepts yield meaningful medical criteria. The missing person cannot be clinically examined, the soul cannot be anatomically located, and though married outside of the caste, one can still pass stringent clinical examinations which would satisfy any life insurance proposal.

Scientific and biological concepts of death refer to the clinical status of the organism. It is important to distinguish between two such concepts, as failure to do so has yielded much talking at cross purposes in recent discussions on criteria for death:

[1] First, there is 'death of the whole organism'. Criteria fulfilling this concept are culturally unacceptable. It would imply destruction of *every* component part, which would mean that putrefaction was the only sure sign of death.

[2] Second, there is 'death of the organism as a whole.' Modern concepts have to be biologically selective, seeking to establish the 'death of the organism as a whole'; that is, seeking criteria to show 'irreversible loss of function of the organism as a whole'. When these criteria are met, then the organism can no longer function as an independent biological unit, and can neither cope with its external environment nor its internal milieu. In its most basic sense this should include irreversible loss of consciousness and irreversible loss of the capacity to breathe. It should be stressed that both the traditional cardio-respiratory concepts and neurological concepts based on brain death yield criteria for death *before* all tissues and component systems have died. A concept, so formulated in terms of 'loss of function of the organism as a whole' recognises that components do not die simultaneously, and that certain functions, such as muscular contractions or spinal chord reflexes, may persist, and that tissues such as skin, bone, or arterial wall might remain viable for a day or two. It is simply that certain transiently persisting functions are not given any significance. In the 18th century, the English clergyman, Stephen Hales, repeated an experiment conducted by Leonardo da Vinci when he tied a ligature to a frog and cut off its head then observed its heartbeat for a short while. Even 30 hours later it was still possible to produce movement in the limbs when stimulated. Babinski recorded, during the 19th century, how individuals executed on the guillotine might retain knee

jerk reaction for up to 20 minutes after decapitation. None of these functions were considered as meaningful exceptions to death of the 'organism as a whole.'

Quite clearly, some concepts of death are more relevant than others. Before assessing competing formulations of the concept of death in current debates it is important to consider guidelines for their assessment.

Irreversibility

Any valid concept of death must be linked to an irreversible physical change in the state of the organism as a whole which can be clearly and unambiguously determined by empirical means. It therefore follows that if a patient were to recover after being pronounced dead, it should not be said that he or she was dead but is now alive again, but rather that he or she was alive all the time, but mistakenly diagnosed. Given the potential reversibility of states associated with the traditional cardio-respiratory concept, then only a brain-centred concept yields necessary and sufficient criteria for death of the organism as a whole. Although circulation has for centuries been considered a point of no return, providing acceptable criteria for loss of function of the organism as a whole, it does not amount to a meaningful concept of death. This is because cessation of heartbeat and circulation is only lethal if it lasts long enough to cause critical centres in the brainstem to die. If the heart function can be replaced in time then death will not occur. But the brainstem is irreplaceable in the way that the cardiac pump is not. This is not a newly discovered fact, but simply a rearrangement of what has been long known. Brain death is neither a new concept of death nor is it an alternative one. It is a reformulation of the traditional concept according to which loss of heartbeat and circulation is not a state of death itself, but an indication (in certain cases where ventilatory support is absent) of the imminence of death. That is to say, according to a brain-related concept of death, criteria such as loss of respiration, heartbeat and pulse, acquire a different status. They are indicators regarding the state of the brain. For the patient is alive until the brain is dead. Physiologically speaking all death is - and always has been - brainstem death. A person may either die from circulatory arrest of sufficient duration to destroy the brainstem or succumb to catastrophes within the head which irreversibly damage the brainstem.

Independence of criteria

It is important both scientifically and morally to maintain a clear separation between the concept and criteria for death and other extraneous factors. Matters relating to the cost of therapy, the anxiety of the family, relatively poor prognosis, quality of life, arguments for euthanasia, and the need to procure transplant organs, are all fundamentally distinct from criteria for death. A meaningful concept of death must be related exclusively to the condition of the patient. For this reason practices have evolved whereby the determination of death is performed by physicians whose primary interest is that of the patient, and wholly distinct from the interests of the transplant surgeons. It is also important that evidence for the 'poor quality of life' of certain brain damaged patients should be discussed only in the context of proposals for therapy options, not in the context of diagnosing death itself. Thus in cases where massive anoxic damage to the cerebral hemispheres leaving most of the brainstem intact results in a persistent vegetative state, a diagnosis of death would be inappropriate despite poor quality of life, as respiratory and other biological functions would remain. Moreover, proposals to classify anencephalic infants or certain categories of persistent vegetative states as dead, despite evidence of brainstem function, for the purpose of organ procurement, reveal fundamental confusions between the definition of death and extraneous needs.[1]

Some opponents of brain related concepts of death have wrongly confused criteria for brain death with policies for the procurement of transplant organs. To avoid this confusion statutes or guidelines for brain death should specify that the primary interest in formulating a definition of death is to recognise a morally significant boundary between the duties owed to dying patient and those which are appropriate to a corpse. For this reason it is mistaken to propose either looser or more stringent criteria for diagnosing the death of organ donors, as this would entail the absurd suggestion that there is a special kind of death for organ donors. The need for a definition of death is a by-product of medical science, and objective criteria for death are essential for the cessation of therapy, and also authorisation of organ removal. But the primary reason in formulating objective criteria must be strictly limited to the interests of the dying patient.

Sophisticated technology in I.C.U.'s has forced upon us a need to understand and define death. If transplant surgery were outlawed, if a supply of artificial organs eliminated all need for human cadaver organs, there would still be a need for precise guidelines on brain death. Although some opponents of brain death criteria have suggested that an interest in procuring transplant organs put pressure on neurologists and neurosurgeons to redefine death, this was not the case. In the 1950's, long before cardiac transplantation, when renal transplantation was still highly experimental, there were profound ethical discussions concerning the value of 'ventilation to asystole' when treatment for patients in irreversible apnoeic coma was obviously hopeless, and increasingly gruesome. These anxieties lay behind Pope Pius XII's[2] remarks concerning decisions to withold resuscitation in cases of irreversible apnoeic coma.

Formulations of brain death: higher brain and whole brain

The majority of countries equipped with sophisticated I.C.U. facilities have adopted some form of a brain death definition, which recognizes the essential irreplaceable nature of brain function. What has to be addressed now is the precise formulation of the concept of brain death. This is not simply a technical matter to be left to physicians; it is a matter of profound philosophical and moral concern.

Higher brain formulations

Higher brain formulations have been proposed by some philosophers and physicians in the U.S.A. and U.K. They are sometimes referred to as 'ontological' definitions of death [3,1,4] according to which death is determined with reference to irreversible loss of structures, such as the cortex and cerebral hemispheres, which are associated with continuing consciousness and cognition. It is argued that since loss of higher brain functions entails loss of continuous mental processes, then an ontological formulation of brain death rests on criteria for loss of personal identity. Ontological definitions do not attribute any signficance to the persistence of other functions, such as spontaneous respiration and heartbeat.

It should be stressed that so far no health authority has adopted guidelines for death based on loss of higher brain functions alone. Among the objections to higher brain formulations is the argument that they lack diagnostic certainty.[5,6,7] Diagnosis and prognosis of the persistent vegetative state, for example, is indeterminate and may take many months before certainty is achieved. There is no widely accepted definition of consciousness, and if there were its absence could not be guaranteed whilst brainstem function persisted. There are moral and aesthetic objections to the treatment of warm, pulsating and spontaneously respiring beings with functioning brainstems, as cadavers fit for organ removal and burial. Furthermore, concepts of continuous personal identity do not yield objective testable physiological criteria, as criteria for personal identity are heavily influenced by culturally based perceptions of what constitutes a person. After centuries of discussion philosophers have so far failed to achieve agreement as to what constitutes personal identity.

Other versions of higher brain formulations of death are based on the view that patients in P.V.S. are incapable of social exchanges and of initiating morally relevant action.[8] This may be so, but it would seem that these issues are related to judgements concerning the quality of life, not its presence or absence, and consequently belong with discussions on therapy options for living patients with severe head injuries.

The whole brain formulation

This formulation finds expression in the USA's Uniform Declaration of Death Act which was adopted by Congress following a recommendation for the President's Commission in 1981.[9] It requires the existence of a state characterised by the 'irreversible cessation of *all* functions of the entire brain, including the brainstem'. In practice, this can never be ascertained by any particular test or combination of tests.

There is a clumsiness in this formulation which is indicative of conceptual uncertainty. It specifies the whole brain whilst needlessly specifying one of its parts, the brainstem. It is not clear exactly which, or how many functions of the brain will satisfy the demand for 'all functions'. How many brain cells have to cease functioning before one can speak of whole brain death? Should it be 100%, 95% or 90%? The Commission's commitment to the whole brain formulation appeared to be an extra-cautious measure. But in this very caution it sought to achieve what cannot be achieved. It is not possible, in the context of suspected brain death, that loss of cerebellar or thalamic function can be directly demonstrated. Moreover, signs of residual electrical activity in isolated neuronal aggregates in the higher regions of the brain do not indicate persistent functioning of the brain as a whole, or the organism as a whole.

The brainstem formulation: the whole brain or the brain as a whole?

Despite the foregoing criticism of the wording in the Uniform Declaration of Death Act, the distinction between the 'whole brain' and the 'brainstem' definition of death is, both philosophically and in practice, of minor significance. In both cases what matters most is the role of the brainstem. Whilst the past twenty years have seen the gradual acceptance of the proposition that death of the brain yields both necessary and sufficient criteria for death of the organism as a whole, the last ten years has seen a parallel development: the gradual realization that death of the brainstem is a necessary and sufficient condition for the death of the brain as a whole - and that brainstem death is therefore itself synonymous with the death of the organism as a whole. The brain is the 'critical system' of the living organism; the brainstem is the 'critical system' of the brain. Brainstem death signifies death of the brain as a whole, not death of the whole brain. From this standpoint residual cellular activity does not indicate the persistence of life.

Brainstem death has been described by a leading neurologist as the 'physiological core of brain death'[10] and consequently loss of function of the organism as a whole. It must be stressed, however, that the term 'brainstem death' does not refer to pathological changes which are confined to the brainstem.

Although brainstem death may very occasionally occur as a primary event it is, in the vast majority of cases, the result of massive increase in intracranial pressure which produces the crucial clinical signs (apnoeic coma and absent brainstem reflexes) which are detected in an ICU when brain death is diagnosed.

The significance of brainstem criteria can be appreciated with reference to its contribution to the continuous function of the organism as a whole. In its upper part the brainstem contains crucial centres for generating the capacity for consciousness. Thus whilst extensive damage to the cortex, from trauma or anoxia, may not cause permanent unconsciousness, there is one functional unity without whose activity consciousness cannot exist. This is the ascending reticular activating system, or ARAS, which is a function of the upper part of the brainstem. Acute, strategically situated bilateral lesions in the paramedian tegmental area of the rostral brainstem entail loss of the capacity for consciousness. In the lower part of the brainstem are mechanisms which control the respiratory centre. Thus lesions of critical areas in the lower part of the brainstem are associated with the permanent cessation of the ability to breathe, which in turn deprives the heart and cerebral hemispheres of oxygen, causing them to cease functioning.

Since irreversible loss of brainstem function necessarily involves loss of *both* the capacity for consciousness and of the capacity to breathe, a very strong case could be made for linking brainstem death to traditional philosophical and religious-based definitions of death as the 'departure of the soul from the body' and the 'loss of the breath of life'. In a very important sense brainstem death is not a new definition of death, nor is it an alternative definition. (It is absurd to speak of alternative deaths; there are many ways of dying but only one way of being dead.) When properly understood, irreversible loss of brainstem function has always provided the mechanism of death.

A diagnosis of brainstem death has two important implications. The first is that the heart will inevitably stop within a very short period. This is an empirically validated observation to which no exception has been found despite thousands of detailed observations of patients ventilated to asystole. The second implication is philosophical in that quite independent of the cardiac prognosis an individual with a dead brainstem is already dead, incapable of responding to the environment, irreversibly unconscious and irreversibly apnoeic.

The essential preconditions and tests for a dead brainstem have evolved over the past twenty years and have been fully discussed by the British neurologist Christopher Pallis[11] who recommends a clinical rather than an instrumental approach. Tests involving EEG measurements are not considered essential to the determination of brainstem death despite their popularity in the media. A flat, or more correctly, an isoelectric EEG reading, is of no significance in the determination of brain death. Patients with electro-cerebral silence have been known to recover and even walk out of hospital within 48 hours of the initial injury. On the other hand, an EEG reading can be obtained from a decapitated head or even a slice of brain tissue in a Petrie dish. The diagnosis of a dead brainstem involves three steps:

1. Preconditions

Ascertaining that the patient is comatose and on a ventilator, and that a diagnosis of the responsible condition is 'fully documented and unequivocally accurate'. There must also be irremediable structural brain damage.

2. Exclusions

The exclusion of reversible causes of a a non-functioning brainstem, among which are hypothermia and drug intoxication.

3. Timing

The conduction of tests, *at the right time*, to ascertain that the comatose patient is genuinely apnoeic with absent brainstem reflexes. This entails an 'unhurried' approach with an attitude that ventilation of the comatose patient should continue 'for as long as it takes to ascertain that the preconditions have been met, and that all conditions that have to be excluded have in fact been properly excluded'.[11] This will vary from case to case, and may be between 6 to 24 hours or more. When it is deemed appropriate to conduct tests, the objective is to [a] ascertain absence of all brainstem reflexes and [b] rigorously document apnoea. This entails a battery of clinical tests (each reinforcing the information to be derived from the others), such that the determination of death does not depend on a single procedure or on the assessment of a single function. For this reason it is mistaken to perceive brainstem death as the loss of a particular organ. Details of the tests have been well documented by Pallis and others.[5,6,9,10,11]

Informing relatives

It must be stressed that disconnection of the ventilator following tests which demonstrate a dead brainstem should not be equated with 'disconnecting life support'. It should be communicated to the relatives that in such circumstances doctors are not giving up therapy and allowing the patient to die, but rather they are ceasing to perform useless treatment upon someone who is already dead.

When transplantation is envisaged the tests should not be carried out by doctors associated with the intended recipient. But when death has been determined, it should be made clear that further ventilation is simply for the purpose of organ viability, not life extension. According to one American lawyer relatives should not be approached for the purpose of granting permission for organ donation until brainstem death has been determined.[12] The point to emphasise is that organ donation should not be perceived as a means of accelerating the moment of death.

Notes

1. Gervais, Karen Grandstrand, *Redefining Death*, New Haven: Yale, 1987.
2. Pope Pius XII, 'The Prolongation of Life', in *The Pope Speaks,* London: Catholic Truth Society, Summer, 1957.
3. Green, M.B. and Wikler,D., 'Brain Death and Personal Identity', in M. Cohen, T. Nagel and T. Scanlon (eds), *Medicine and Moral Philosophy,* Princeton: Princeton University Press, *1981, 49-77.*
4. Zaner, R.M.(ed), *Death: Beyond Whole-Brain Criteria,* Dordrecht: Kluwer, 1988.
5. Lamb, D., *Death, Brain Death and Ethics,* London: Routledge, 1985.
6. Lamb, D., *Organ Transplants and Ethics,* London: Routledge, 1990.
7. Lynn, Joanne, 'The Determination of Death', *Annals of Internal Medicine,* 99, 1983, 264-6.
8. Gillett, G.R., 'Why Let People Die?' *Journal of Medical Ethics* 12., 1986 pp.83-6.
9. President's Commission, President's Commission for the Study of Ethical Problems in Medicine and Biomedical and Behavioural Research, *Defining Death*, Washington: US Government Printing Office, 1981.
10. Pallis, C., *The ABC of Brainstem Death*, London: BMJ, 1983.
11. Pallis, C., 'Brainstem Death' ch.19. *Handbook of Clinical Neurology*, Vol 13. (57): *Head Injury*, ed. R. Breakman, 1990, pp.441-496.
12. Annas, G.J., 'Brain Death and Organ Donation: You *Can* Have One Without The Other', *Hastings Center Report* June/July, 1988, pp.28-30.

Notes on authors

Richard F. Kitchener Ph.D., is Professor and Chair at Colorado State University, USA. He obtained his Ph.D at the University of Minnesota in 1970. Professor Kitchener is interested primarily in problems in the philosophy of science, especially Psychology. He is the editor of *New Ideas in Psychology*. His earlier work includes: *Piaget's Theory of Knowledge*, 1986. He is the editor of *The World View of Contemporary Physics: Does it Need New Metaphysics?*, 1988. His most recent interests include Psychological Epistemology. Dick enjoys skiing, back-packing and singing barbershop quartet.

Mark B. Woodhouse Ph.D., is Associate Professor of Philosophy at Georgia State University where he teaches courses in metaphysics, parapsychology, and Eastern thought. Author of the widely adopted text, *A Preface to Philosophy*, he also serves as editorial advisor for the *Journal of Near Death Studies*. He is currently completing a manuscript on new paradigm thought, *Worldviews in Transition*.

Joseph Wayne-Smith, Ph.D, is Research Fellow in Philosophy and Sociology at Flinders University, South Australia. Among his publications are *Essays on Ultimate Question*, Gower, 1988; *The progress and Rationality of Philosophy as a Cognitive Enterprise*, Gower, 1988; and *The High Tech Fix*, Gower 1990. He is a regular contributor to *Explorations in Knowledge*

Robert C. Trundle, Ph.D., is currently Professor at Northern Kentucky University, where he teaches, among other courses, Logic and Existentialism.

He has recently been honoured as Outstanding Junior Faculty Member of the College of Arts and Sciences, Northern Kentucky University, for teaching performance and scholarly achievement. The latter reflects numerous publications on comparative philosophy (Camus and Nagarjuna), philosophy of science, philosophy of religion, existential phenomenology, and political as well as axiological philosophy. Among his earlier publications is *Beyond Absurdity,* UPA, 1986, which was co-authored with Professor R. Puligandla (Ph.d, MS. physics).

Jean Marie Trouvé is Lecturer of Science Communication at the University of Poitiers, France. His major work is *A Theory of Scientific Knowledge*, with publications in French, Italian and English in various areas of science studies. In 1990 he held a visiting appointment at the Research Unit on Science and Technology at the University of Rome 'La Sapienza'. He has contributed several articles to *Explorations in Knowledge*.

Kevin White BA, Dip. Soc Sci., Ph.D., a graduate of Flinders University of South Australia, has lectured in Sociology at Flinders, in the Department of Science and Technology Studies (specialising in the social history of medicine) at Wollongong University, New South Wales; and is currently Lecturer in Sociology at Victoria University, Wellington, New Zealand. He has contributed several articles and reviews to *Explorations in Knowledge*.

Lucy Frith BA, M.Phil. is currently researching a Ph.D. thesis on Bioethics and the issues surrounding the new reproductive technologies. Among her publications is *Society, Dichotomies and Resolutions: An Inquiry Into Social Synthesis,* Gower 1992. She is a regular reviewer for *Radical Philosophy* and *Explorations in Knowledge*. Her recent teaching assignments include Philosophy at Coleg Harlech, Wales; Medical Ethics and Philosophy at St Martin's College, Lancaster, and an Applied Ethics course for the Open University, England. Lucy plays a good game of cricket.

Brian Goodwin BA, BSc. MSc. Ph.D. is Professor of Biology in the Open University. He has research interests in Biochemistry and developmental aspects of biology, especially in rethinking of evolution via the relationship between ontogony and phylogeny. His books include *Temporal Organization in_Cells*, 1963, *Analytical Physiology of Cells and Developing Organisms*, 1976. He is co-editor of several collections including *Theoretical Biology: Epigenetic_and Evolutionary Order From Complex Systems,* 1989.

Gerry Webster BSc. Ph.D., teaches in the School of Biological Sciences, University of Sussex, and is Chairperson of the Human Sciences Organising Committee. His research interests include the Philosophy of Biology, especially in relation to the problem of biological form. He is currently engaged in a conceptual critique of the 'evolutionary paradigm' from a realist and non-reductionist perspective. He is co-author with Brian Goodwin of *Il Problema della Forma in Biologia*, Armando Editore, Rome, 1988.

E.K. Ledermann, MD, FRC Psych, was Consultant Physician at the Royal London Homeopathic Hospital and a Consultant Psychiatrist at the Marlborough Day Hospital, England. Since retiring from the National Health Service he has continued his private practice in Harley Street, London. His books include *Philosophy in Medicine*, Gower, 1986, *Existential Neurosis*, 1972, *Good Health Through Natural Therapy*, 1976, and *Mental Health and Human Conscience*, Gower, 1984. He contributes regular articles to *Explorations in Knowledge*.

David Lamb BA. Ph.D. is Senior Lecturer in Philosophy at the University of Manchester and General Editor of the Avebury Series for Gower Publishers. His research interests include Philosophy of Science, Philosophy of Medicine and Philosophical Problems in Space Exploration. His publications include *Discovery, Creativity and Problem-Solving,* Gower, 1991. He jointly edits *Explorations in Knowledge* with Dr Susan M. Easton and Professor Gonzalo Munevar. He also trains dogs for water rescue work and restores old buildings.

Index

Aristotle, 38,81

behaviorism, 144
Bell, J.S., 28-30
Beloff, J., 50-1
Bergson, H., 51
Bijke, W.E., 87-8, 110
Bohm, D., 30, 32
Braithwaite, R., 4
Brodbeck, M., 5,20
Bunge, M., 118

Carnap, R., 4-7, 9,10,13-17,21
Cartwright, N., 79
Chalmers, A., 60,70,71
Compton,J., 73,75

Dawkins, R., 131
death, 40-43, 178-186
Douglas, M., 122
dualism, 34-35,42-44,121-3

Einstein, A., 27, 74
EPR effect, 28-30,43Feigl, H. 4, 9, 10,15

Feigl, H. 4,9,10,15
Feyerabend, P.K., 19, 62, 70,82,85,118,116
Fleck, L., 120-1
Florman, S.C., 76
Foucault, M., 115,123
Frassen, C. Bas. van., 79

Goodman, N., 7

Hacking, I., 79
Heidegger, M., 5,61,76,77,84
Hesse, M., 118
Hobbes, T., 129
holism, 57,169,172
holograms, 31-33
Hubner, K., 79,82
Hume, D., 70,77,79

Kant, I., 67,69,70,72-3,75,77,81
Kockelmans, T., 19,20,25
Kuhn, T., 15,17,24,70,82,85,92-4,112, 114,165
Lakatos, I., 165

Laudan, L., 165
Lockley, R.M., 150
logicism, 6,7
Lorber, J., 33

Malcolm, N., 33,35,51,52
Marshall, N., 50,54
McMullin, E., 20
memory, 49-55
Midgeley, M., 127,133,142-150
Mulkay, M., 93,110

Nagel, E., 4
Neurath, O., 5
Newton, I., 74, 82
Newton-Smith, W.H., 66,70-72, 80
Niddith, A.P., 6,21
Noske, B., 147-8

Pallis, C., 184-6
Penfield, W., 51,59
Pinch, T., 87-8, 110
Plato, 52,144
Popper, K., 8,12,19,80-81
positivism, 1,4,6,8,10-12,17,20,21
Pribram, K., 30-35,46

quantum theory, 28, 30

realism, 12,13,29,50,56
Reichenbach, H., 4,5,23
relativism, 72
Rorty, R., 77
Russell, B., 5,6
Ryle, G., 147

Sabom, M., 41
Salmon, W., 4
Sartre, J.P., 75,77,79-80,85
scientism, 19
Sheldrake, R., 37-39,43,47,55-58
Singer, P., 138-141
Smith, M., 162
Soloman, R., 59
Squires, R., 51
Symons, D., 130-2

Taxonomy, 156-7
Tesh, S., 117
Thomas, L., 151

Watson, J.B., 144
Weismann, A., 159-160, 162-4
Wilson, E.O., 129-130, 132-3, 136-142
Woolgar, S., 89, 111
Woosley, S., 74

Zohar, D., 32

HAROLD BRIDGES LIBRARY
S. MARTIN'S COLLEGE
LANCASTER